THE FIVE DEEPS EXPEDITION

ATLANTIC • SOUTHERN • INDIAN • PACIFIC • ARCTIC

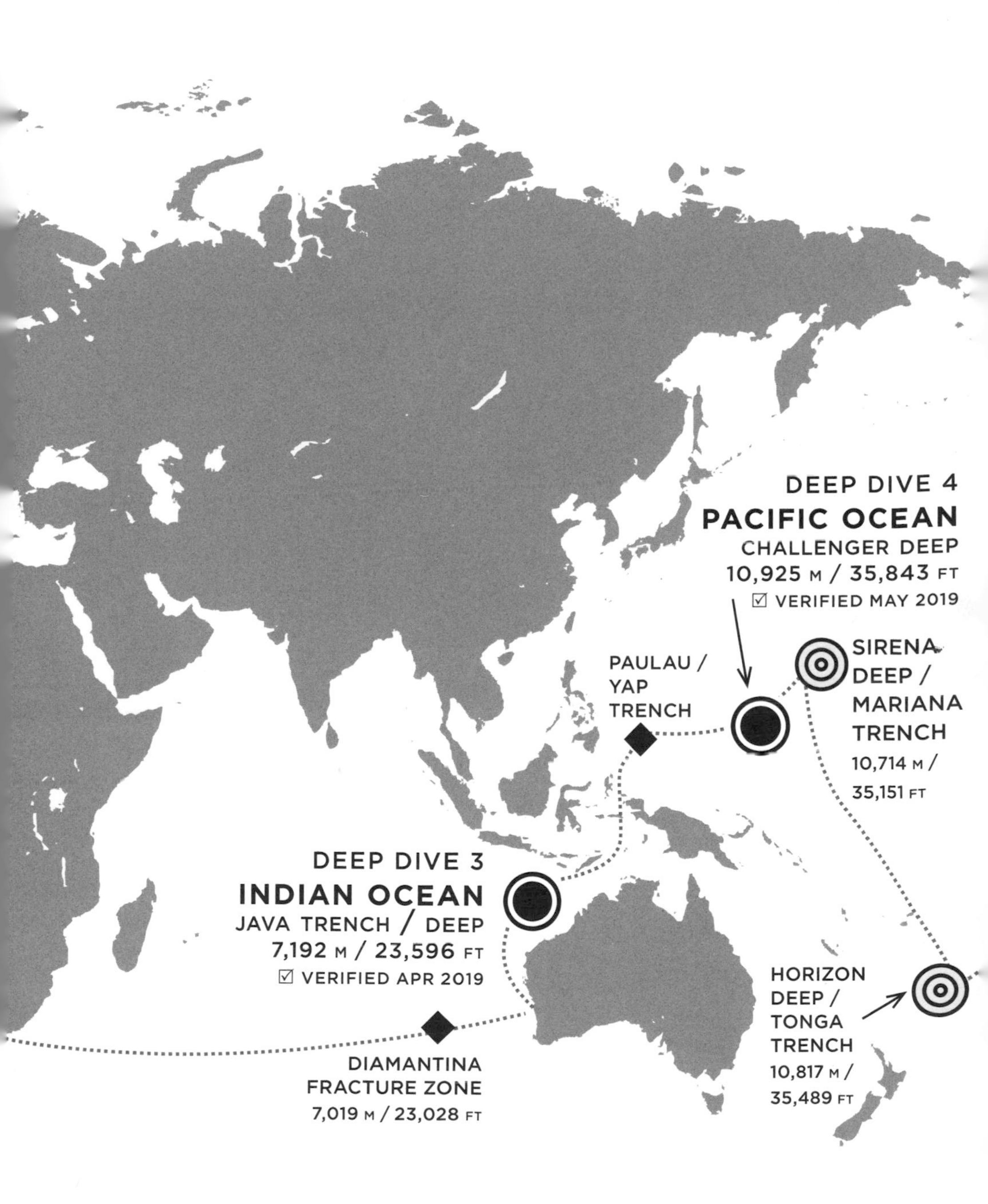

PRIMARY "DEEP" DIVES

SUPPLEMENTAL DIVES

SONAR & MAPPING OPERATIONS

EXPEDITION DEEP OCEAN

EXPEDITION DEEP OCEAN

THE FIRST DESCENT TO THE BOTTOM OF ALL FIVE OF THE WORLD'S OCEANS

JOSH YOUNG

PEGASUS BOOKS
NEW YORK LONDON

EXPEDITION DEEP OCEAN

Pegasus Books, Ltd.
148 W. 37th Street, 13th Floor
New York, NY 10018

First Pegasus Books cloth edition December 2020

Interior design by Maria Fernandez

ISBN: 978-1-64313-676-9

10 9 8 7 6 5 4 3 2 1

Printed in the United States of America
Distributed by Simon & Schuster
www.pegasusbooks.com

For Samuel Macy Harrell III

and

In memory of Sally Bowers Harrell

with eternal gratitude

Knowledge of the oceans is more than a matter of curiosity. Our very survival may hinge upon it.

—President John F. Kennedy,
March 1961

CONTENTS

PROLOGUE
SPACE TO SEA

Space, that final frontier, has long been the ultimate in exploration. In the early 1960s, a space race between the United States and the Soviet Union shifted into high gear. Space travel captured imaginations around the world. Sending humans into the heavens in a rocket and allowing them to look back at Earth was a romantic notion. And then there was the burning question of whether or not there was any life on the Moon.

Three major space programs funded by the U.S. Congress were launched by NASA with varying goals and results. Project Mercury, which began in 1958, was the first human spaceflight program administered by NASA. The program's goal was to put an astronaut into orbit around the Earth. With millions following the developments on radio, that goal was achieved by John Glenn on February 20, 1962, and he became a modern-day hero.

Project Gemini, which ran from 1961 through 1966, took the next step toward landing a man on the Moon. In a capsule built for two astronauts, the Gemini spacecraft flew low orbit missions that were working to perfect Mercury's orbital techniques and prove the feasibility of in-orbit docking, necessary to complete a Moon landing. But the human cost was high, as three astronauts lost their lives during Gemini training. Adding the loss of life to the financial cost, a vigorous debate began over whether space travel was actually necessary or more of a quixotic pursuit.

NASA pushed ahead. The Apollo program was the final step in landing a man on the Moon. The spaceflight program, through many travails and the loss of yet three more lives in Apollo 1, achieved its goal. On July 20, 1969, with the world at a virtual standstill as hundreds of millions watched on television, Neil Armstrong and Buzz Aldrin landed Apollo 11's lunar module, the *Eagle*, on the Moon. Later that day, Armstrong became the first human to walk on the lunar surface.

Each of the space exploration programs built on the previous one, and one could not have been possible without its predecessor. To reach the Moon through these three monumental programs, which adjusted for inflation cost more than $175 billion, there were incremental steps toward the completion of the full mission. The spacecraft had to launch, dock in orbit, and the astronauts had to then land the *Eagle* on the surface of the Moon. Finally, the astronauts needed to return safely to Earth. For all the money, engineering know-how, and technology, the core mission of the programs was fairly straightforward: to safely land a man on the Moon, take a soil sample, and return it to Earth for further study.

As a result of the space race and the continuing fascination with the Moon, 533 people have orbited Earth, and 12 have walked on the Moon. As a species collectively, including the Soviet Union's early missions, we've gone "up" into space with great success. But what about going "down" to the bottom of our oceans? What has come of the "oceans race?"

The answer is next to nothing, because there isn't really one.

Oceans cover some 70 percent of the planet's surface, driving weather, regulating temperature, and ultimately supporting all living organisms; and oceans have also been a vital source of sustenance, transport, commerce, growth, and inspiration, yet humans have explored only 20 percent of this virtual lifeblood. Most people struggle to even *name* all five of the oceans on our planet. Nothing has been more taken for granted—and more feared—than the deepest waters of the world.

The best known oceans, certainly to Americans, are the two that border the continental United States, the Atlantic Ocean to the east and the Pacific Ocean to the west. The largest and deepest ocean of them all, the Pacific Ocean covers some 63 million square miles and touches more than 40 countries and 23 additional territories, stretching across the curves of the globe from the United States to Australia to Thailand.

Further north and east of the Atlantic, past Greenland and continuing to the northern tip of Russia, is the Arctic Ocean, generally identified because of its relationship to the Arctic Circle and renowned for its cold waters and polar ice caps. The Indian Ocean stretches from Asia at the north, to Africa on the west and Australia on the east, and skirts Antarctica to the south. Its warmer temperature cannot support as much sea life as the other oceans.

The most difficult one to name, for most, is the lesser traveled Southern Ocean, which wraps around Antarctica and is usually defined as all the water south of 60 degrees south latitude—known by mariners as the "screaming sixties" because of its common, and ferocious, storms. It's the place of nature documentaries that feature the seals and whales of the Antarctic coastline and the penguins inhabiting the remote South Sandwich Islands far to the east of South America.

Despite the vital connection between the oceans and mankind, as of the turn of the calendar to 2018, only three people on two separate missions had been to the bottom of the Mariana Trench in the Pacific Ocean. Named the Challenger Deep, it has long been believed to be the deepest point of our five oceans at somewhere close to 10,916 meters, which is more than 35,000 feet, or roughly 6.7 miles.

The first mission to accomplish the dive was the U.S. Navy's Project Nekton on January 23, 1960, something of an oceanic Mercury mission with the goal being to reach the bottom. The submersible *Trieste* was piloted by Swiss engineer Jacques Piccard and U.S. Navy lieutenant Don Walsh, *Trieste*'s commander. It proved with brute force that a massive steel sphere, made buoyant by huge tanks that were filled with gasoline that was lighter than water, could reach the bottom of the Challenger Deep.

The second mission occurred on March 26, 2012, when filmmaker James Cameron piloted his single-person, steel-capsuled *Deepsea Challenger* submersible to the bottom. His mission took the next step and showed that not only could we reach the bottom in a manned submersible, but we could take the next step and remain there and explore for hours, making it the Gemini version of deep ocean exploration that built on what the *Trieste* had accomplished.

Those two missions were more than a half-century apart, a stunning commentary on the lack of curiosity about the deepest point on our planet. And no one had even attempted to reach the bottom of the other four oceans, and therefore no one knows what, if anything, of meaning might be there. The fact is that scientists don't fundamentally understand how the oceans interact with the atmosphere or even each other, because we don't have much reliable data—perhaps because 90 percent lies below 1,000 meters. Below 1,000 meters, the water pressure is over 1,400 pounds per square inch and marine engineers say that at that point, "things get really complicated." This depth is also generally referred to as the "crush depth" of most modern military submarines, and any craft capable of going below that depth requires special engineering, construction, and testing.

In 2018, a group set out to build a new machine that could go not just beyond 1,000 meters, but to 11,000, and—finally—dive to the bottom of all the world's oceans, a hoped-for Apollo-like deep ocean mission. To accomplish this, a 46,000-nautical-mile expedition, wrapping around the globe and then some, was set in motion and financed by a ponytailed Texan with a penchant for exploration that had led him to the peak of the highest mountain on each of our seven continents and to ski both poles, completing the so-called Explorers' Grand Slam.

A self-proclaimed loner, polymath, and sci-fi junkie, the private equity investor risked a sizeable portion of his net worth to finance the project and attempt to become the first human to reach the bottom of all five oceans, video the ocean floor, and return with sediment and sea life samples.

The submersible to take him to the depths was designed by a Briton who has never read a book on subs and doesn't believe in formal engineering training as a means of building them. Instead, he relied on a vision of what was possible based on a combination of his imagination and a study of the physical properties of the components. The submersible was built by the Brit's employer, a Florida-based company owned by two partners, one a somewhat combative former gemologist, the other a gregarious Canadian who drops an f-bomb with flair in every third sentence. Certification would need to come from a narrow-focused German expert who knew as much, or more, about submersibles than any of the other principals.

The expedition leader was a former New Zealand park ranger living in Seattle whose firm specializes in exotic sea expeditions and was once the subject of a "Modern Love" column in the *New York Times* in which his wife wrote, "My father warned me about guys like you." The science portion of the mission was slotted to be led by a speaks-his-mind Scot from Newcastle University on his 55th oceanic expedition who had literally written the book on the Hadal Zone—all depths deeper than 6,000 meters—called, quite appropriately, *The Hadal Zone*.

The ship, flying under a Marshall Islands flag to avoid unwanted complexities associated with flying a U.S. flag, was captained by another jovial, but highly professional Scot who earned his sea legs in the oil and gas business, and run by his fellow European engineers and a rotating crew of Filipinos with an Austrian doing the cooking. The entire expedition was filmed for not one but two television events, which were to be produced by an Emmy award–winning documentary film producer known for his ability to produce amazing footage and landmark nonfiction films, but who maintained such a tight grip on all media relating to the expedition that it frequently led to clashes with other team members.

This is the story of how they came together on the first expedition to ever attempt to dive to the bottom of the five oceans, in hopes of creating a marine system capable of reliably, safely, and repeatedly journeying to any point on the bottom of the ocean. Such a thing had never existed before in human history and they were determined to change that.

CHAPTER 1
THE EXPLORATION GENE

On a typically sticky late May evening in 2014, Victor Vescovo guided his Tesla through light traffic in his hometown of Dallas. He was on his way to one of his regular dinners with his older sister, Victoria. His mind was whirring, as his latest voyage of exploration was taking shape. He had been poking around the Internet for weeks, doing research, and was ready to put his new, somewhat radical idea in motion.

Tall and lanky, Vescovo is a native Texan, though he doesn't look much like one. His long face and gray beard are reminiscent of a centuries-old Italian painter. He has graying blond hair that he wears in a ponytail more indicative of a sculptor than a 49-year-old partner in a private equity firm.

The minute Vescovo sat down at a local favorite restaurant in her town of Coppell, ironically named Victor's Wood Grill, his sister knew something was up. The two are close even beyond being on the same page, or finishing each other's sentences. They're on the same wavelength, *and* amplitude.

Once when they were playing Pictionary, Vescovo looked intently at his sister while drawing a single line, perfectly straight and rapidly, like it was shot out of a gun. From his expression and the *way* he drew it, she immediately shouted out, "Laser." It was the right answer, of course, and they weren't ever allowed to be on the same team again.

That night at dinner, she saw that determined, focused look in his eyes that she had seen so many times. When they were children, she had watched him build space capsules out of large cardboard boxes. He would "wire" the capsule to the family car with a household extension cord for power, and outfit it with a communications apparatus made from an old rotary dial telephone. He would then spend hours piloting his "craft" through space, using accurate commands and terminology from books he had read. Remarkably, he also built a working radio from a Radio Shack kit at age eight, and then a computer with cassette-tape-based storage when he was thirteen.

As an adult, Vescovo took his adventurous spirit to new heights, literally, and cemented his place among the elite mountain climbers of the world. He climbed the tallest mountain on each continent, thus being one of the 350 people to complete one of the ultimate climbing challenges known as the "seven summits." All the while, he was on his way to becoming a licensed helicopter pilot and later acquired a fixed-wing jet rating, eventually purchasing both a helicopter and jet so he could fly them himself.

"I've got this new thing I want to do," he said to Victoria. "Going to the bottom of all five oceans. I've researched it and I think the technology is there to build a sub that can make it. I even have a rough idea of the logo"—pausing, he gestured with his right hand—"pen?"

Victoria reached into her purse and pulled out a doctor's appointment card and a pen. A thin, elegant woman with straight silver hair, her distinguishing feature in public that night was a neck brace, which she covered with a scarf. In 2006, she had fallen while rollerblading and broken her neck. After five surgeries and the placement of titanium rods, she was still left with a lifetime of recurrent pain. The two of them together also carried a much deeper pain, the death of their sister, Valerie, who had committed suicide in 2002.

Vescovo sketched the outline of a badge on the card. Inside the badge, he drew five jagged, vertical lines in the shape of repeating V's. The troughs of the V's represented the bottom points of the five oceans, Atlantic, Pacific, Arctic, Indian, and Southern.

"So the seven summits and now the five deeps," he said. "It's almost symmetrical, right? I'm going to try and go to the bottom of the five oceans. Nobody has ever done it."

Victor Lance Vescovo has a unique perspective on life, one that he has defined himself rather than letting others define for him. Like most people, part of it comes from how and where he grew up, the experiences of his youth, and the more formative ones as an adult. Much of it though comes from his internal life, the time he has spent wrapped up in science fiction novels, a massive number of history books, and studying and playing military simulations.

Born in Dallas in February 1966, Vescovo is youngest of three children and only son to an Italian-American father, John Peter Vescovo from Memphis, Tennessee, and a German-Irish mother, Barbara Frances Lance (hence his middle name) from Waycross, Georgia. His father worked for Columbia Records, and the family moved around in Vescovo's early years, from New Canaan, Connecticut, to Silver Spring, Maryland. Living in the northeast erased any trace of a Texas drawl.

As a child, he was voracious reader and an avid fan of science fiction. Jules Verne's novels and *Star Trek* were early favorites. In his teens, he discovered Frank Herbert's landmark epic *Dune*, Isaac Asimov's *Robot* and *Foundation* series, the cyberpunk works of William Gibson, and David Brin's aqua-themed *Startide Rising*. Vescovo also discovered Iain Banks's genre-breaking Culture series in 2010. With ten books, the series was like the gift that kept on giving for him.

"I read everything I could get my hands on as a child and during my school years," he says. "I would actually just sit and read the encyclopedia—that was fun for me. I wanted to see the world, understand it, and do interesting things. And don't even get me started on maps. I could stare at maps for hours."

Intellectually, he was advanced. Some early IQ testing in grade school showed that he was off the charts, and school administrators told his parents that he needed to be suitably challenged. He was eventually enrolled in St. Mark's School of Texas, an all-boys prep school in Dallas. Reserved and introverted, he loved chess, military war games, and Dungeons & Dragons, and was a voracious reader of military history. He discovered books about exploration and eventually learned that Dick Bass, a St. Mark's alum, had summited each of the highest mountains on the seven continents. This planted a seed in his mind that perhaps, if his body and skill permitted it, he might be able to replicate the feat.

He attended college at Stanford University. His freshman year, he learned to fly a small Cessna, and a love of aviation was born. Known by his family and friends as Lance throughout his childhood, when he graduated from Stanford and started his first job in San Francisco, he asked his friends and family to start calling him by his first name, Victor.

"Victor had more gravitas than Lance," his childhood friend Matt Lipton says. "I think the change said to him that he wanted to be taken seriously."

At the time, he was considering going into the military. He had discussed this with his Stanford foreign policy professor, Condoleezza Rice, who later became president of the university and then national security adviser and secretary of state under President George W. Bush. She encouraged him to consider becoming a reservist while continuing on his business career—since it was obvious he loved both potential careers—advice that he eventually acted on.

Taking just three years to finish his undergraduate work, he graduated from Stanford in 1987 with a double major in political science and economics, with distinction. By his own admission, he did nothing but study, fly, and practice martial arts during his college years. Developing a social life held little interest for him.

Over the next dozen years, his life was dominated by three pursuits: a varied work life, serving in the Navy Reserve, and climbing the highest mountains.

Before Vescovo started his first job, at Bain & Company in San Francisco, the 21-year-old traveled to Africa with money he had saved up from trading stocks on the side. He ended up in northern Tanzania at the base of Mount Kilimanjaro. Upon seeing it, the only thought that occurred to him was: "Oh yeah, I really have to climb that."

Showing up at the gate with some money and absolutely no experience, he met another young man with a similar desire. They joined forces on the spot, found a guide and porters, and began climbing. By the time they reached the lip of the volcano several days later, and only a few hundred meters from the true summit, Vescovo was extremely sick. His body wasn't accustomed to the heights, he had a bad gastrointestinal bug, and the mountain had taken a punishing physical toll on his untrained body. He was forced to return to the bottom.

But in that first climb, he was bitten—hard—by the climbing bug. "I really liked the adventure part of it, of not knowing what was going to happen next, of would we make it around the next bend," he recalls. "I vowed I'd go home, train properly, work out very hard, and come back and reach the summit." A full decade later he did return, and climbed to the summit and back in just three days versus the normal six or seven.

While working at Bain & Company in San Francisco from 1987 to 1988, he was offered a PhD scholarship at the Massachusetts Institute of Technology (MIT) to study defense analysis. He matriculated at MIT but ended up pursuing and earning a master's degree in Political Science in just one year. His focus was on conventional military warfare analysis, statistics, and operations research. He wrote a master's thesis titled "The Balance of Air Power in Central Europe: A Quantitative and Qualitative Analysis"—a series of increasingly complex mathematical simulations of theater-level air warfare in central Europe that determined, based on an analysis of relative pilot training, doctrine, and the weapons each country had, who would have won an air war.

He attributes much of his academic success, as well as his later success in business, to his faith in math. He often says, "I may not have deep faith in a lot of things, like religion, but I do have deep faith in math." It's one of the maxims that guides his life, both in his investment ventures and in his exploration.

"There's a great line from the movie *Margin Call* where they are talking to a young analyst who has a PhD and he explains what his thesis was about," he related. "His boss says to him, 'So . . . you're a rocket scientist.' He says, 'I was.' They ask, 'Then why are you here on Wall Street?' The guy replied: 'It's all just numbers really, you're just changing what you're adding up.'"

From 1989 into 1990, Vescovo returned to Dallas and moved in with his father, who lived alone after his divorce years earlier. He took a humdrum job at a large telephone company to earn some money and recover from what had been a mentally exhausting experience at MIT. At the phone company, however, he came to understand the 9-to-5 mentality and the different priorities people had in their lives—like family above all—and usually doing only what was asked.

"I heard water cooler conversations about 'how can I make this new initiative go away,' 'how can we stretch out this work,' and 'this is the way we've always done things, so stop rocking the boat by trying to make it better,'" he recalls. "I learned how organizations are *really* resistant to change, though at the time I really didn't know how a direct experience of how people 'on the line' often thought would help me when I later worked to turn around such companies. People on the line just have very different priorities from management and owners. Success requires that you understand people and their priorities even if they are different from your own."

He moved into the world of high finance by taking a job at Lehman Brothers in New York in 1990 to 1991, during the *Liar's Poker* era. While working for Lehman, Vescovo was contacted by a Navy recruiter who told him about a special program for people proficient in foreign languages. He pursued the program, and following Condi Rice's advice, joined the U.S. Navy as a reserve intelligence officer and, in 1993, was commissioned as an ensign.

"I felt that joining the reserves was a way to participate in something bigger than myself," he says. "And I was absolutely fascinated with everything about the military."

❧

Climbing fast became part of who he was, as well as his elixir in life. At age twenty-five, in 1991, he climbed Mount Elbrus in Russia. He went on a commercial expedition with nine people, but was one of only three that made it to the summit. While the group was on the mountain, there was a coup in Moscow, and they had to wait a few days before returning to Moscow. It was a tense time. "When we left Moscow, the hammer and sickle flag was flying, but when we came back, it was the Russian flag," he says. "The world literally changed while we were on the mountain."

In 1992, Vescovo was transferred to Saudi Arabia for a year, where he worked for the Saudi Ministry of Defense in the Economic Offset Projects Office analyzing business ventures proposed to them by foreign contractors. There, he learned Arabic and traveled throughout the Middle East. He returned to the U.S. to attend Harvard Business School, from 1992 to 1994, again living the life of "an intense nerd," as he sums it up, and focusing on high-order finance and learning how to value technology. He graduated in the top 5 percent of his class and was named a Baker Scholar.

His next climbing challenge came in 1993 when he attempted to summit Aconcagua in Argentina, the highest mountain in South America. A few hundred feet from the top, he fell and was badly injured. His team was ascending, unroped. Vescovo put his foot on a boulder, jarring it and causing a rockslide. He slid down the mountain with rocks bouncing around him and was knocked unconscious for a minute. When he came to, he couldn't speak and had excruciating lower back pain where a boulder had impacted his spine, and his right leg was partially paralyzed.

The group was too high for a helicopter rescue so a team of four French climbers and his own team helped him down to base camp. Suffering from hypothermia, he was stabilized and then medevacked to a hospital the following day. He ended up spending months in physical therapy and ultimately made a full recovery.

"Victor is always pushing himself to ultimate limits," his sister says. "Being close to death is something that brings life into focus for him. It's how he finds meaning."

His work life took a turn in 1994 when he returned to Bain & Company and formed the most important relationship of his business life, with Ted Beneski, the co-head of Bain's Dallas office. Beneski gave Vescovo the most complex deals. In a two-year period, Vescovo worked on the Continental Airlines turnaround, reliability process improvement at Dell Computer, and the merger integration of global aerospace giants Lockheed and Martin Marietta.

In his time away from work he focused on training and climbing. Two years after his fall on Aconcagua, he returned to Argentina and reached the summit of the mountain on his second attempt. A year later, he spent three weeks climbing Denali, also known as Mount McKinley, the highest peak in North America.

During this period, Vescovo was assigned to "Top Gun" on the Naval Reserve side. For four weeks a year on and off, he taught intelligence officers how to brief pilots on combat operations. He went on active duty for half a year in 1996, serving on the USS *Nimitz* in the Persian Gulf, and on the USS *Blue Ridge*. In 1999, he was mobilized for the Kosovo War in Bosnia, where he was directly involved in planning and assessing the effectiveness of NATO's bombing campaign against Serbia.

His military service was sent into high gear on September 11, 2001, in the wake of the terrorist attacks on U.S. soil. That night, Vescovo received a phone call that he was being mobilized again. "Within the span of a week after 9/11, my girlfriend and I broke up, I left my job, sent my dog to live with my sister, moved out of my apartment and was on plane to Pearl Harbor," he recalls. "I pretty much ended up in a vault and didn't come up for air for fifteen months."

He was made a Naval Intelligence officer, and was sent to the Joint Intelligence Center, Pacific, located at Pearl Harbor, to work on counter-terrorism analysis in the Pacific rim, primarily in the Philippines, Bangladesh, and Indonesia, based on his knowledge of Islamic culture and fluency in Arabic. This stint on active duty lasted fifteen months, and ended in December 2002. Over the course of the next five years, he would return to work in counterterrorism from time to time. During his time overseas, for fun, he wrote a book titled *The Atlas of World Statistics*, a fairly esoteric mashup of statistics and cartography. After his active duty stint ended, he climbed Mount Vinson in Antarctica, its highest peak.

It was when Victor was serving in Pearl Harbor that he received the most tragic news of his life: his sister, Valerie, had committed suicide. The middle child in the family, Valerie worked as an MD, licensed as an anesthesiologist in Portland, Oregon. It was known that she had battled with acute depression since her teenage years and was pushing through a rough period in her life, but like so many cases, it wasn't known just how serious hers was. Vescovo received emergency leave to attend the funeral.

While he doesn't hide what happened, like most personal issues it's not a subject he is willing to spend much time delving into. The death of his sister, however, served as a continual reminder that life is short and that he must take every experience to its fullest point. Like many before him, Victor coped by turning his full attention to work.

His next business venture would become his most significant, one that drove his net worth well beyond $100 million and allowed him to buy a six-seat Embraer Phenom jet and a Eurocopter 120 helicopter, both of which he pilots. It came in 2002 when he started the Dallas-based private equity firm Insight Equity as a founding partner.

Formed by Ted Beneski, Ross Gatlin, and Vescovo, Insight Equity specializes in turnarounds and business enhancement for mid-market industrial companies with revenues usually around $100 million. The firm has raised more than $1.4 billion in four capital raises and deployed it in a wide variety of industries, including heavy construction, energy, defense, industrial pollution control, and electronics. At the time Vescovo began contemplating diving the five oceans, he was serving as interim CEO of one company and chairman of four others.

"I guess I've become pretty good at time management, that, and I can type really fast, which helps," he says. "I'm not much for vacations either."

And there was climbing to be done. In 2008, he first attempted to summit Mount Everest but made it only halfway up. Battling frostbite on his fingers, he descended. "I also wasn't psychologically ready," he says. "My stepmom was undergoing chemotherapy and other things in my life were filling my head, and it just didn't feel right. You don't climb that mountain when you're distracted and it doesn't *feel* right."

He returned to Mount Everest in 2010. After a month and a half of climbing, on the final push to the summit, a clear weather forecast turned into a snowstorm. The guides and the Sherpa huddled together to make the call whether to go for the summit or return to the base. The weather window was closing and it was now or never for this climb. They had plenty of oxygen and the climbers were strong, so the decision was made to go. Victor's personal Sherpa on summit day was none other than Kami Rita Sherpa, who now holds the record for most summits of Everest in history, at twenty-four. That gave him no small degree of confidence, and the storm cleared almost all other climbers from the mountain, giving his team an all-important clear shot to the top.

On May 28, 2010, at 8:20 A.M., Victor reached the summit. Because of the poor weather and the threat of the storm worsening, his group spent all of fifteen minutes at the highest point on the planet. The hike down to the helicopter to take them off the mountain lasted three and a half days.

"In mountain climbing, it's not all about reaching the top, it's also about coming back down. It doesn't count if you don't come back," Vescovo often says.

The following year, he completed the "seven summits" by climbing Carstensz Pyramid in Irina Java, Indonesia, the highest mountain in Australia/Oceania. He then went on to ski the North and South Poles, making him the thirty-eighth person to complete the Explorers' Grand Slam.

He traces his philosophy of climbing and exploration to Roald Amundsen, the Norwegian explorer of the polar regions in the late 1800s and early 1900s, who was the type of explorer Vescovo admires and aspires to be. Amundsen was very methodical in his preparations and left as little as possible to chance. This was very much in contrast to Robert Falcon Scott, who led a separate expedition to the South Pole five weeks after Amundsen. Because of poor planning, Scott and his team that went with him all died.

"If the epitome of talent in war is to win without fighting, the height of skill in exploration is to achieve your objective with no heroics needed," Vescovo says. "Amundsen said—more or less—that adventure is a sign of bad planning. He *worked the problems* so that when he made the trek, he had the correct solutions to the problems. He did that through extensive preparation, realistic assumptions, thorough planning, and ruthless discipline in execution. Heroism is exciting to read about—planning and preparation aren't. But when you are actually the one doing the exploring, believe me, you want a no-drama expedition."

Owing partly to the danger of climbing, his side career in the Navy, and an almost obsessive dedication to his work, Vescovo has never married and has no children. He also points out that his family's matrimonial track record is poor. His mother was married four times and thrice-divorced, his father had two marriages and one divorce, his older sister the same, and his younger sister one divorce. Even all of his uncles, aunts, and cousins are divorced without exception. Adding to this, his best friend from high school has married four times. Vescovo figured his odds were not good.

In 2010, Vescovo met Monika Allajbeu. A native of Albania, the statuesque Allajbeu has the features of a model with the warmth of personality of a girl who grew up in a large family. She often calls Vescovo "Lance," his middle name used during his childhood. A year after they began dating, she moved into Vescovo's house in north Dallas, and made plans to enter dental school. She prides herself on giving Vescovo space to do what he wants.

Never marrying has also allowed Vescovo not to feel responsible to anyone, such as children or a wife, should something go horribly wrong during his extreme adventures. As he points out, he has nearly died on a mountain twice. He also adds: "There's too much to do to be tied down. And, maybe, I think I would lose that edge that pushes you to your limit."

Part of Vescovo's drive to explore dangerous pursuits may be explained by what Dr. Glenn Singleman, an MD, refers to as the Exploration Gene Theory, which helps explain why certain people need to seek out thrilling pursuits to feel fully alive. Singleman, who holds a world record in base-jumping, has studied this phenomenon and served on several exploration expeditions.

The theory is rooted in research done by Dr. Marvin Zuckerman of the University of Delaware. He found that on the DRD4 gene on chromosome 11, people have a different number of copies, between two and eleven. The gene itself codes for dopamine receptors in the central nervous system and measures their sensitivity.

"If a person has two copies of the DRD4 gene, they have sensitive dopamine receptors in their central nervous system," Singleman explains. "The importance of that is that dopamine is the 'feel good' neurotransmitter of the central nervous system. Every time that person feels a WOW moment, that is dopamine in action, the feel of being really invigorated. So if you have sensitive dopamine receptors, you don't need much stimulation to get your dopamine receptors activated. That's if you have two copies. If you have eleven copies of the DRD4 gene, you have insensitive dopamine receptors in the central nervous system. That means to get that buzz, you have to have high sensations."

Singleman says these sensation-seekers tend to be the race car drivers, test pilots, skydivers, and mountain climbers. They are also people with a greater propensity for other dangers, including criminal behavior.

"The brains of these people are wired different," Singleman says. "People like Victor have to undertake these extreme adventures to get that dopamine release. They also respond differently to trouble and are able to overcome a fear response when faced with extreme danger. Rather than being overwhelmed, they are able to reorient themselves, respond, and solve problems in life-threatening situations."

Singleman takes things a step further, saying that people like Vescovo are necessary to push the boundaries of what is possible. "To get into the possibility of the human condition, you have to have those 'test pilots,' like astronauts, who go out and test the boundaries," Singleman says. "They are essential to the health of society because otherwise we would be stuck in place on so many levels."

Contemplating the existential question of "why attempt to dive the five oceans?" Vescovo acknowledges the need within himself to undertake extreme adventures. "There's a certain group of people, genetically or otherwise, where it's a compulsion to attempt difficult things," he says. "Something inside drives us, whether it's psychosis or the exploration gene. To *not* do these things is *very* difficult for us because you feel you are not properly living within the constructs of your own personality. We feel like the horse in the cage before the race—raging against the steel constraints confining us.

"Building on that theme," he continues, "this adventure has the added enhancement to do something completely new, a lure all of its own. Is there a bit of egotism involved? Absolutely. But that's not the driving force."

So when Vescovo was completing the seven summits, he was trying to stave off a letdown by thinking about the nature of exploration, which had come into focus in the news as private missions to space were being planned by billionaires Sir Richard Branson, Elon Musk, and Jeff Bezos. He needed a new challenge, a unique challenge that no one had achieved. The media was again talking about space, but what about the depths? Who had gone to the deepest points on the planet, to the bottom of the five oceans?

When he learned that no one had, it was time, Vescovo thought, for someone to try. For him, this pursuit would combine the thrill of climbing with the lessons he had learned in business, as it also presented major technological and organizational challenges. This would also call on skills from his naval background, both in terms of exploring the seas and the type of military-style planning that would be required. What started as a matter of curiosity fast became a goal, and he decided that he was at least going to find out if he had the wherewithal to finance such an undertaking, and if there were people who could help him get it done.

"I also found it objectionable that here we were in the year 2014—2014!—and we still have not been to the bottom of all five of our oceans," he says. "In some respects, it comes down to that. Someone should have done this. I have the resources to try, and it sounds like a fun adventure. So I started looking seriously into just what would it take to accomplish that goal."

Sitting across from Vescovo at dinner that night in Dallas, his sister Victoria was convinced that he would somehow make this dream a reality. "When Victor sets a goal and gets that look in his eyes, you know that is going to happen," his sister says.

It's no understatement saying that she held the minority belief for quite some time.

CHAPTER 2
THE SEARCH FOR A SUB

If curiosity is the driver of all exploration, it has been in limited supply when it comes to deep ocean diving and the vessels necessary to reach extreme depths. The deep diving submersible world was a fairly small community with a limited track record, as Victor Vescovo discovered in 2014 when he began researching how he could reach the bottom of the five oceans, and who had attempted—and succeeded—in reaching the ocean floor in the past.

The first vessel to reach the bottom of what was believed to be the deepest point of the oceans was a bathyscaphe (meaning "deep ship") named the *Trieste* after the Italian city where it was built. Piloted by U.S. Navy lieutenant Don Walsh and Swiss engineer Jacques Piccard, the son of the bathyscaphe's designer, the *Trieste* reached the bottom of the Challenger Deep in 1960. The Challenger Deep is a long gash in the Mariana Trench in the Pacific Ocean, off the coast of Guam. The *Trieste's* depthometer measured water pressure and plotted it on a paper cylinder. It recorded—after some technical corrections—a final revised depth of 10,912 meters (35,800 feet).

The *Trieste* was a fairly basic craft that worked much like a hot-air balloon—only it did so underwater. (Though often called submarines, bathyscaphes are not technically submarines because they are built like hot-air balloons and have limited mobility. A submarine is a self-contained submerging vessel in which the crew can live for prolonged periods of time. This is in contrast to a submersible, which cannot be deployed as an independent vehicle, and requires a surface support ship.) In fact, the *Trieste* was designed by the Swiss scientist Auguste Piccard, who was known for his record-breaking helium balloon flights before turning his attention to bathyscaphes. Piccard applied the principles of ballooning to the design of his deep-sea craft.

The *Trieste*, which weighed some 150 tons, was composed of a steel sphere large enough for two aquanauts and had a single acrylic window from which the pilots could view their surroundings. The average hull thickness of the sphere was seven inches, the same as its Plexiglas viewport. On top of the sphere were several tanks holding 34,200 gallons of aviation fuel. The concept was that the gasoline in the tanks was lighter than water and can't be compressed, thereby providing buoyancy to the heavy craft. To send the bathyscaphe down the water column, large air tanks on its top were filled with water to initiate descent. Two large tubs on the bottom each contained eight tons of steel shots that would be released at the ocean's floor for the craft to regain buoyancy and ascend.

On the *Trieste*'s Challenger Deep dive, while passing through 9,000 meters, a secondary window that was in the entrance tube to the sphere, but was not a pressure boundary, cracked. Walsh and Piccard looked at each other in a moment of severe anxiety. But when they realized they weren't dead, they shared a nervous laugh and continued their descent. The extreme pressure had compressed the cracked acrylic and sealed it from leaking.

In its first and only dive in the Challenger Deep, the *Trieste* took 4 hours and 48 minutes to reach the bottom, stayed just 20 minutes because its landing impaired visibility, and then took 3 hours and 15 minutes to return to the surface.

Part of the reason it went down only that one time was that the Navy found out that after one of the *Trieste*'s test dives, the sphere had shifted slightly, and epoxy glue was applied before the Challenger Deep dive. When Walsh later sat down with a Navy admiral about future dives, he lectured Walsh, "In the Navy, we don't glue the hull of a sub." Walsh also says that Piccard told the brass at Office of Naval Research, who were funding the program, that he did not think the sphere was safe for full ocean depth dives.

The *Trieste*, however, was refitted several times and used in many subsequent government underwater missions. It was retired in 1964, and in 1980 was placed in the U.S. Navy Museum in Washington, D.C.

The French Navy built the next major deep-diving submersible in 1961, the *Archimède*, their second bathyscaphe. Like the *Trieste*, it too used gasoline for buoyancy. In this case, the gasoline used was 42,000 gallons of hexane, the lightest gasoline available.

One main difference between the *Trieste* and the *Archimède* was in the sphere that housed the pilots. The *Trieste* had three pieces of steel forged together—and glued together, as mentioned—while the *Archimède* pilot capsule was one welded piece of steel. The result was that the *Archimède* was far stronger. The Trieste had a safety factor of 1.1 at full ocean depth, but the added ability to withstand pressure gave the *Archimède* a factor of 3.

In its history, the *Archimède* made multiple deep ocean dives. In 1962, it reached a depth of 9,560 meters (31,350 feet) across multiple dives in the northwestern Pacific Ocean. Then, in 1964, the *Archimède* dived to what was believed to be the deepest point in the Atlantic Ocean in the Puerto Rico Trench, though the location turned out to be incorrect. There was never a mandate to dive and explore the deepest points of the other oceans—these were exclusively scientific, not record-breaking, missions.

More advanced manned submersibles were built in the ensuing years, but progress was slow, at best. The years of intermittent deep-diving activity in the late 1950s began to temper as media, public, and government attention drifted to the "space race."

The *Alvin*, a two-pilot, one-passenger sub, was built in 1964 by the U.S. Office of Naval Research and operated by the Woods Hole Oceanographic Institution (WHOI). The cost was pegged at $1 million. The 23-foot, 17-ton sub, which had two robotic arms, made headlines for taking Robert Ballard to the *Titanic* wreck in 1986 and for exploring the Deepwater Horizon oil spill. The sub has made more than 4,400 dives conducting research, diving to depths of 4,500 meters, and is still in use today.

The advantage of the *Alvin* over its predecessors was that it was far lighter. The lightness was created by using syntactic foam for buoyancy rather than the gasoline tanks used in the *Trieste* and the *Archimède*. Syntactic foam is composed of millions of hollow glass microspheres wrapped in an epoxy resin, and it is strong enough to withstand the pressure at deep depths while remaining buoyant.

"We would have eagerly adopted foam and titanium if they were available to us," Walsh says. "The bathyscaphes and their huge quantities of highly flammable hydrocarbons were essentially large bombs, a constant source of concern to all of us. But that's all we had."

Four *Alvin*-class submersibles were subsequently built, and they were continuously upgraded since their original launch, with depth capability increasing and productivity enhanced with evolving modern systems. Still, none was even remotely capable of diving to the oceans' deepest points.

In 1984, the French built a follow-up to the *Archimède* named the *Nautile*, operated by a new French government ocean agency. Also called the SM-97, meaning it was capable of surveying 97 percent of the planet's ocean waters, the *Nautile* could dive to 6,000 meters.

The interior diameter of the *Nautile* was the same as the *Archimède's*, but its sphere was constructed of titanium, making it far lighter, and therefore easier to launch and recover. It was used to dive to the *Titanic* wreck multiple times and also to search for the black boxes from the downed Air France flight 447, an Airbus 380 that crashed into the Atlantic Ocean on June 1, 2009.

The *Mir*, a three-person submersible, came next in 1987. It was similar to the *Nautile* in many regards. Capable of reaching 6,000 meters, the *Mir 1* and its updated version, the *Mir 2*, were also used several times to dive the *Titanic* wreck, at 3,800 meters, which fast become a benchmark of respectability in the deep submersible vehicle community. The ship that carried the submersible, the *Akademik Mstislav Keldysh* (or "*Keldish*" for short in Western nomenclature), had a more sophisticated launch and recovery system than the *Alvin*, allowing it to dive in rougher seas. However, the *Mir*'s drawback was that anyone using it for expeditions had to supply their own cameras and other science-related equipment.

The technological bar in deep ocean submersibles was advanced by the Japanese government with the development of *Shinkai*, a three-person submersible put into service in 1989 that could dive to 6,500 meters. It was owned and run by the Japan Agency for Marine-Earth Science and Technology, which received funding from a consortium of large companies. Consequently, in its time, the *Shinkai* was outfitted with the most advanced electronics, the highest resolution cameras, the most accurate sonar, and high capacity batteries that allowed for longer dives. Its support ship, the *Yokosuka*, also had a multibeam sonar to measure the diving depths.

In 2010, the Chinese government built the deepest diving submersible since the *Trieste*, the *Jialong*, which reached a maximum depth of 7,062 meters (23,169 feet) in the Mariana Trench in 2012. Two years later, when Vescovo was doing his research, it seemed to still be operational, though no one was sure of its true depth capabilities or what it was doing.

❧

The common factor in almost all submersibles built and operated after 1964—until 2012—was that they were limited to diving in the top two thirds of the ocean. But they could not explore the absolute deeps.

The top 200 meters of the oceans is known as the Epipelagic Zone, where common sea life exists. From 200 to 1,000 meters is the Twilight Zone, which is just beyond the reach of sunlight and home to bioluminescent species. The Midnight Zone stretches from 1,000 to 4,000 meters and favors sea creatures with slow metabolic rates. This is followed by the Abyssal Zone, from 4,000 meters to 6,000 meters, an area so deep that it is unaffected by surface weather. The Hadal Zone starts at 6,000 meters and goes down to the bottom of the ocean at more than 10,900 meters, and this is where it becomes difficult for life.

The name "Hadal" is derived from the Greek god of the underworld, Hades, as the ocean depths are often regarded as dark and scary as hell itself. The deepest parts of the Hadal Zone exist in the large ocean trenches that are knife-like canyons in the seafloor. These were formed by tectonic activity as the seafloor crustal plates converge, at what are called convergence zones, with one pushing under the other, a process called subduction. This occurs at continental boundaries when the more dense seafloor plate pushes under a relatively buoyant continental plate. The frictional forces result in significant terrestrial volcanic activity. Exploration of the Hadal Zone in manned submersibles has been virtually nonexistent, as scientists have relied on remotely operated vehicles, or ROVs.

Until 2012, of all the active submersibles, only *Jialong* could reliably dive into the Hadal Zone—and it appeared to do that rarely. That year, the bar was raised when a submersible capable of reaching full ocean depth was constructed.

Named the *Deepsea Challenger*, the submersible was designed by Australian Ron Allum for an expedition led by filmmaker James Cameron and financed by National Geographic, Rolex, and other partners. Though the sub was tested in a pressure chamber at Penn State University before diving, it was a prototype and was not certified by any international organization or depth-rated for commercial use. Its cost, according to published reports and those on Cameron's mission team, was in the $10- to $12-million range, not including several million dollars spent on the expedition.

The "Kawasaki racing-green" submersible was built for practicality, and not aesthetics. It was 24 feet long and only 43 inches wide, so small inside that the pilot had to squeeze his body into the sphere and lie flat rather than sit up. The space was so constrained that he could not fully extend his arms.

The sub's steel capsule was 2.5 inches thick and weighed 11.8 tons. It had twelve exterior thrusters that allowed the pilot, using joysticks, to "drive" the sub just above the seafloor. The two keys to the sub's full ocean depth capability were its syntactic foam created by Allum and called Isofloat, which was used for buoyancy, and powerful lithium batteries with rapid recharging systems. The sub was outfitted with 3D cameras and had a mechanical arm for picking up sediment and rock samples that could be deposited into a retractable compartment on the sub.

After four tests dives in Sydney Harbor and off the coast of New Britain in the Pacific, Cameron went for the bottom of the Challenger Deep. On March 26, 2012, Cameron successfully piloted the submersible to the bottom of the eastern portion of the Challenger Deep, making it only the second mission to reach the nadir, and the first solo mission.

Weather and time constraints forced Cameron to dive at night. He spent two and a half hours exploring the bottom. When he returned to the surface, the support ship had difficulty locating him. The ship called on billionaire Paul Allen's yacht, which was nearby, to help guide it to Cameron's position and recover him.

During the dive, the sub encountered a series of problems, many of which were caused by an accelerated schedule that resulted in the use of what turned out to be some subpar parts. Many of the Chinese-made thrusters reportedly failed, and there were multiple system failures, though none life-threatening. Though plans were made to repair it, the submersible never dived again.

Vescovo was surprised by his research. In all, there were only five working submersibles capable of going below 7,500 meters. All five were owned and operated by governments. Since the *Mirs* went out of service in dives below a few hundred meters in 2005, there wasn't any submersible that could reach the *Titanic*, at about 3,900 meters. Only two missions, fifty-two years apart, had reached the bottom of the Challenger Deep in the Pacific Ocean, and *no one* had reached the bottom of the other four oceans. Moreover, no one had really tried. How was that even possible, he asked himself.

More research revealed that scientists didn't even definitively *know* where the deepest points of the Atlantic, Arctic, Indian, and Southern Oceans were because they hadn't been properly mapped with modern technology. The bathymetric maps of the oceans that existed were inconsistent. Early bathymetric study of the oceans, which involved lowering cables to measure the depth, were wildly inaccurate. Updated methods of using an echosounder, or sonar, to ping a beam of sound to the seafloor were more reliable, but a lack of interest or funding had led to incomplete data across vast stretches of all five oceans.

Why had no government gone to the bottom of the oceans? Certainly, the U.S. government wouldn't likely fund such a mission, and even if they did, it would take years, be mired in red tape and appropriations and cost far more than it needed, he reasoned. "What is an elephant?" he liked to joke to people. "It's a mouse built to government specifications."

He kept digging. Had any private investor tried?

Sort of, was the answer.

Sir Richard Branson, the billionaire owner of Virgin Atlantic, announced a mission to dive the five oceans, but that had turned out to be more sizzle than substance. Virgin Oceanic's *DeepFlight Challenger* submarine was unveiled in April 2011, with Branson describing its mission to dive to the bottom of all five oceans as "the last great challenge for humans," something Vescovo agreed with.

The *DeepFlight Challenger* was designed in 2000 by Graham Hawkes, a well-known British sub builder, for the adventurer Steve Fossett to make the first solo dive to the bottom of the Challenger Deep. But when Fossett died in a plane crash in 2007, the project was put on hold. It was subsequently purchased by investor Chris Welsh, who then partnered with Branson.

Designed to look like an airplane with a large front window, the single-pilot sub was about 24 feet long and had a 10-foot wingspan. Initial pressure testing showed that the sub would likely only be able to do a single full ocean depth dive, but even that proved to be optimistic. The vessel's domed viewport showed signs of cracking during testing. The pressure hull, made of carbon fiber, also would not be able to withstand sustained pressure on repeat dives to full ocean depth.

Though the company's website promised "five dives, five oceans, two years, one epic adventure," the sub made only five shallow test dives, during which it encountered a series of insurmountable problems. In 2014, Branson put the entire project on ice, but didn't rule out returning to it. "Starting a new venture takes a 'screw it, let's do it' attitude and finding the right partners to help us achieve the unthinkable," he said at the time. "However, business is also about knowing when to change tack.

We are still highly passionate about exploring the bottom of the ocean. However, we are now widening the focus of the project and looking for new technology to help us explore the ocean and democratize access at reduced cost and increased safety."

Since then, there have been incomplete reports of an ostensibly private Chinese venture, called the "Rainbow Fish Project," to build a full ocean depth submersible and custom support ship. But the best information indicated that the project was millions of dollars over budget and its completion date kept being pushed back. And besides, Vescovo wasn't sure the Chinese would be excited about loaning their craft to a former U.S. Naval Intelligence officer.

It was time, Vescovo thought instead, to build his very own reliable submersible to send him where no one had gone before.

The first option for a submersible was to purchase the *Deepsea Challenger.* "Logically, I thought that if that sub still exists and given that it had gone to the bottom of the Challenger Deep, then it could go to the bottom of all the oceans," he says.

Vescovo also reasoned that it would be less expensive to buy an existing sub and update it than to have one built. He found that after its success diving the Challenger Deep, the *Deepsea Challenger* had been donated to the Woods Hole Oceanographic Institution on Cape Cod.

In the spring of 2014, Vescovo contacted WHOI, and explained that he was interested in purchasing the *Deepsea Challenger.* As it wasn't a normal inquiry, and perhaps not entirely believable, a few months passed before he received a response. That July, the WHOI special projects manager, David Gallo, invited Vescovo to see the sub.

Vescovo flew to Boston, rented a car and drove to Cape Cod. The two met at a coffee shop to discuss Vescovo's intentions, and then headed over to see the *Deepsea Challenger.* Vescovo was very surprised at what he saw.

The history-making sub was crammed into a corner in the back of a large warehouse, sort of a metaphor for where deep ocean diving had gone since Cameron's mission. It looked worn out. There were cables hanging out of the hull, and several of the thrusters appeared to be barely attached. Gallo explained that the batteries and the thrusters would need to be completely replaced, along with a very long litany of other items.

"It was almost like he was discouraging me from taking it over," Vescovo recalls. "He kept calling it a prototype and saying it needed a full-on rebuild. What I was also surprised to see was just how small the interior was in real life. It was basically a small bubble that the pilot has to be shoehorned into."

Vescovo left Woods Hole wondering what a refit would cost even if he chose to pursue it, and shifting his thinking toward having a sub built from scratch. Still, he followed up with Woods Hole a few more times, but he never heard back.

The fate of the *Deepsea Challenger* would later take a bizarre turn. In July 2015, the sub was loaded onto a flatbed truck en route to a container ship that would transport it to Australia for an exhibition. Driving on Interstate 95 near Stonington, Connecticut, the transport truck's brakes malfunctioned and ignited the rear wheels, catching the sub on fire. A part of the sub's bright green hull was charred. Later, the sub was relocated to California for refurbishment with reports that it was eventually headed for a maritime museum in Australia.

~

As it became apparent that the *Deepsea Challenger* would not be the best option, or even a viable one, Vescovo turned his attention to finding a company that could build a full ocean depth submersible. He happened upon the website for Triton Submarines, a small company in Vero Beach, Florida, that appeared to be a well-regarded submersible manufacturer. His interest was sparked by Triton's affiliation with Ron Allum, the designer of the *Deepsea Challenger*.

He was particularly encouraged that the company's website actually advertised a full ocean depth submersible, called a "36,000/3"—meaning a submersible that could dive to 36,000 feet with 3 people inside—though there was little information about it. That was exactly what was needed.

On September 14, 2014, Vescovo emailed Triton Submarines' head of marketing and sales, Mark Deppe. He expressed interest in leasing a Triton 36,000/3 if Triton had one available.

"To be quite direct, my intention is to partner with a firm and support team to journey not just to the Challenger Deep, but the other deepest points of the world's four other oceans," he wrote. "This has never been done, and I would very much like to expend the resources and time to be the first to do so."

He also mentioned that the dream to dive the ocean was an extension of his recently completed quest to climb the seven summits. And he indicated that he was the cofounder of a successful private equity firm and had the financial resources to see such a deep ocean diving project through.

A response quickly came back, and a call was set up. It turned out, however, Triton had a *vision* for a full ocean depth sub, but they hadn't actually *built* one. No client had stepped up and ordered it, so all that existed was a colorful drawing and a dream of what might be possible. And no one was quite sure how much it would cost. Undeterred, Vescovo decided he would meet the Triton team and find out just what the possibilities were.

CHAPTER 3
THE SUB BUILDERS

At the time Victor Vescovo contacted Triton Submarines in 2014, the company was housed in a nondescript, warehouse-style building off the beaten path, down the street from a Burger King in Vero Beach. The manufacturing floor resembled a marine version of the Wright Brothers' bicycle shop, with spare parts strewn out over workbenches and the floor littered with circuit boards, propellers, and widgets all shapes and sizes for the subs in progress. The communal offices were in a two-story attachment to the warehouse. The thin, wood-paneled walls caused employees to joke about having to wear headphones for privacy. Behind the main building was a cement courtyard used for assembly space, and at the rear of a property was a second manufacturing building.

The company was formed in 2007 by L. Bruce Jones and Patrick Lahey. Jones and Lahey met when Jones was running U.S. Submarines, which he did from 1987 through 2007. Lahey worked for Jones on several projects and then the two decided to form a company to concentrate on building small submersibles.

"I gave Patrick half the business with the understanding that I would not be there full-time and that he would run the day-to-day operations," Jones explains. Though Triton was headquartered in Florida, Jones continued to reside in rural Idaho. So Lahey became the operational leader of the company, Jones its marketer.

The two men are Type A personalities and share many things in common, but have very different styles in business.

Jones, 63, is a bear of a man with a round face that suits his jolly demeanor. He punctuates his stories, good and bad, with a hearty chuckle. He's not shy about expressing his role in the company's projects. Though this has led to his reputation as a self-promoter, he characterizes it as being an advocate for Triton. He often takes a very direct approach that can rub clients the wrong way, something he says can be necessary to protect his team and keep things moving forward.

Lahey, seven years his junior, has spiked white hair and tanned skin from spending much of the last thirty-six years on ship decks deploying subs. He casually drops f-bombs, using them positively, negatively, and angrily in equal doses. Appearing a bit high-strung to some, he carries the weight of the company's projects on his shoulders, literally walking with them pitched forward, but he shines a beacon of optimism on every situation. He takes a back seat to the client and operates under the premise that the client is always right—or if the client isn't, he's at least very deferential to their view.

Both men came to love the ocean at an early age. Jones grew up all over the world in mostly port cities, living in Hong Kong, Manila, Singapore, and Jakarta, as well as on oil rigs for periods of time. In total, he has visited 121 countries and lived in 20. His mother was a civil engineer, his father a marine engineer. His grandfather also made his living in the maritime industry, owning a large marine fleet construction company and, according to Jones, had a role in inventing containerized shipping.

Jones learned to scuba dive at age eight in the Persian Gulf, triggering an immediate fascination for underwater exploration. When he was in the eighth grade and living in Singapore, his parents were displeased with the school choices so they sent him to boarding school in the U.S. at Culver Military Academy in northern Indiana. "I hated it, but I appreciated what it did for me," he says. "And I learned to fly airplanes there."

He later attended Trinity University and after graduating went to work in gemology. The interest continues to this day, as he keeps what he calls the world's most advanced private gemological lab on the second floor of his Idaho home.

After his first marriage ended in divorce, Jones met his current wife of thirty-two years, Liz. The two shared the same adventuresome spirit. When they met, Liz had just finished sailing around the world. They now regularly go on trips in their small plane, ride in their hot-air balloon, and take excursions in their motor home.

Jones is a *bon vivant* who will sip champagne in the afternoon if the mood suits him. He is full of colorful stories, many of which sound slightly exaggerated or even apocryphal. To wit: "When we were treasure hunting in the Philippines, I made a recovery of $300 million of gold bullion, but [President Ferdinand] Marcos was my partner and he took it all, and we got escorted to the airport at gunpoint."

Lahey's love affair with the oceans also started in his childhood. He spent his early years in landlocked Ottawa, Canada, but when he was seven, his father moved the family to Barbados for three years. Lahey's new backyard was the crystalline waters of the Atlantic Ocean.

"The ocean made an indelible impression on me," Lahey recalls. "Looking at it every day solidified my desire to have a life of the ocean. There's a great line from Jacques Cousteau: 'The sea, once it casts its spell, holds one in its net of wonder forever.' I'm one who found himself under that spell."

Lahey got into the sub game in his late teens. He had become a certified scuba diver at thirteen, and by the time he turned eighteen, he was working as a commercial diver in the oil and gas industry.

At that time, the industry used manned submersibles for exploration. The submersibles were more like oversized, wearable diving suits. The pilot slid inside a pressure-controlled tube that barely had enough room for him to move his arms to operate the controls.

Lahey made his first solo dive in one of those subs at age twenty-one off the coast of Santa Barbara, California, to inspect an oil rig at 1,400 feet. "I vividly remember the cobalt blue water all around me and the sense of liberation I had," he says. "As a diver, I was accustomed to spending time in a decompression chamber when I came up. In a sub, you don't have to do that. I decided at the moment I surfaced that subs would be the focus of my professional life."

∞

Triton started small and experienced growing pains in a very niche business. The first two Triton submersibles, which were built when Jones and Lahey were still at U.S. Submarines, were 1,000/2s, meaning two people could occupy the capsule and dive to a maximum depth of 1,000 feet. The company then graduated to building acrylic-sphered 3,300/3s that look like aquatic spaceships. These became the company's bread and butter.

Triton's first three years were "dismal with small growth," as Jones characterizes them. The company was manufacturing one to two subs a year. It had persistent issues with suppliers, such as when the UK company that built the acrylic spheres for its 3,300/3s delivered ones that were yellowish and of dubious durability. The company actually managed to destroy two spheres. Triton switched to a German manufacturer, but there was a long research and development cycle that slowed Triton's output before the company arrived at a new method to make optically clear and stronger acrylic spheres.

By 2011, the pace increased and Triton's staff had grown from the original three full-timers to thirty-six. Revenues were also growing, at a rate of 25 percent per year. High net worth clients began ordering Triton subs, such as billionaire investor Ray Dalio, who owns two. By 2014, when Vescovo contacted Triton about building a full ocean depth submersible, the company had increased its output to two or three subs a year and had made twelve.

"Triton has always been a little bit of a Skunk Works," Lahey says, referring to Lockheed Martin's legendary development shop that coined the term for innovative undertakings that operate outside research norms. "It's an unusual place full of eccentric people, but people who are passionate about what they do and who infuse their energy into every part they make."

When Jones first read Vescovo's query, he wasn't sure what to make of a Texan who said he wanted to fund a deep diving sub. Jones knew everyone in the sub world, but he had never heard of Vescovo. "Patrick and I live on a steady diet of rejection," he says. "We've been through it all. Every wacko that wants a submarine over the last thirty years, we've heard from them. But Victor was a very credible guy from the beginning and pretty easy to deal with."

Out of left field, as the two sides were just beginning a correspondence about the sub, Jones pitched Vescovo on a marijuana business investment for cannabis oil production in California and Washington. "My response was, 'Um . . . okay . . . no thanks,'" Vescovo says. "A bit of a yellow flag, too."

Based on Vescovo's strong interest in pursuing the full ocean depth sub, Triton began internally gathering information on what it would take to build it. The company was focused on its dream to build a sub with a transparent pressure boundary, essentially a glass sphere. The properties of glass are such that a perfect sphere could sustain the pressure, but there were a variety of drawbacks that came to light.

While the right type of high-pressure glass could bear the compressional stress, there could not be any type of glass-to-metallic interface. This meant the glass sphere would have to be a clamshell design with all the equipment necessary to power and operate the sub located inside the sphere. A further problem was that the glass company Triton had approached was proving unreliable.

Vescovo wasn't sold on the idea of a glass sphere for three passengers. "I went back to them and said, no, let's go tried and true," he says. "I don't want to sink a ton of money in bleeding-edge tech that may or may not work."

Triton was forced to abandon the idea and focus on something more practical. "There were obvious advantages to glass, but as Victor so eloquently put it, he wasn't interested in funding a science experiment," Lahey says. "And fair enough. It was a shrewd and fairly wise decision on his part because at the end of the day, as exciting as a glass pressure boundary is at full ocean depth, it's still far off and unknown."

During this time, on May 20, 2015, Triton invited Vescovo to the Bahamas to dive in one of their three-person submersibles, a Triton 3,300/3 (meaning a maximum depth of 3,300 feet with three passengers). Lahey had two Emirati clients on the ship. The two men were chatting away in Arabic when Victor walked in—and started talking to them in Arabic. Lahey was floored.

Vescovo had never been inside a sub, and the experience cemented his interest in pursuing a full ocean depth sub.

In the Bahamas, Vescovo also met the company's lead sub designer, John Ramsay, who was based in the UK. Ramsay, who is in his late thirties, has a boyish spirit about him, tousled hair, a thin scruff of facial hair, and eyes that open extra wide when he emphasizes a point.

Ramsay had started at Triton as a consultant in 2006 and went full time in 2012 after designing sixteen subs for Triton. Previously, he had worked in the UK building military-oriented submarine rescue systems, in which a small sub with a decompression chamber attached to it would dive to a larger submarine to bring people safely back to the surface.

Ramsay would be tasked with coming up with a workable design for the full ocean depth submersible. As complex as the undertaking would be, Ramsay operates on a fairly straightforward philosophy of "if you can't explain it, you don't understand it."

"One of the objectives of that dive in the Bahamas was to prove to Victor the importance of being able to see out of the submarine with your own eyes and not just through cameras," Ramsay says. "At the time he wasn't interested in all the specialized requirements of a submersible, but we wanted him to see them."

Vescovo was immediately impressed with Ramsay. "He struck me as a child prodigy who grew up and got to build things he had only dreamed of," Vescovo says.

∞

After the Bahamas dive, the discussions became more specific. Some basic questions had to be answered before Ramsay could begin the actual design. First among them was how many passengers the sub would carry.

Vescovo wanted to dive solo to the bottom of the deeps, and he wasn't convinced that having viewports was worth the expense or the risk. No full ocean depth submarine had ever had more than one viewport, and Triton wanted to put in *three*. Instead, he suggested they put several high-definition cameras on the outside to film the depths and show the pilot what was outside—referring to them as "virtual windows." He reasoned that a single-pilot sub would also be far roomier. What he was looking for was an elevator to the bottom of the ocean and back to the surface. "Let's just start from just the basics: put me in a metal ball, and drop me to the bottom," he only half-joked to Jones.

Both Jones and Lahey pushed back. They argued that a two-person sub with viewports had far more utility and ongoing value after Vescovo finished his expedition. "If you can't see out of the sub, what's the point?" Lahey said. If the sub held two people, the pilot and a passenger (such as a scientist or a filmmaker), they could dive and interact together. Finding a buyer for a two-person submersible with viewports would also be easier than selling a one-seater that you couldn't directly see out of. Triton also wanted to put a mechanical arm on the sub to collect sediment and rock samples.

"I talked to Patrick and I said, 'Victor, respectfully we are not going to build a one-person sub with no viewports,'" Jones says. "I told him, 'One, you will never be able to sell it, and two, you are not going to be able to do any science. You have the risk of this being perceived by the world as just doing an ego-motivated record attempt with a submarine that will have no future value.'"

As Jones pointed out to Vescovo, for different reasons, that was what had happened with the two previous deepest diving subs. Both the two-person *Trieste* and one-seater *Deepsea Challenger* only completed one dive to the Challenger Deep. While both were intended for multiple repeat dives to the Challenger Deep, neither sub attempted a second dive to full ocean depth, and the *Deepsea Challenger* ended up never diving again and was donated to a museum rather than sold.

"We convinced him to build a sub that would move the needle," Jones says. "We wanted to build a sub, not just for Victor, but for others to follow in his footsteps and continue ocean exploration. We wanted a sub that was fully accredited by an independent, third party agency that would attest to its capabilities."

There were only a handful of companies that certified submarines, and none had ever even been asked to certify one to full ocean depth.

James Cameron and his team had pressure-tested the *Deepsea Challenger*, but it had not gone through an independent certification process, thereby leaving it classified as an experimental prototype.

The cost to have the sub certified would be in the range of $500,000, not including the pressurized testing of all components, which could be twice that much. Nevertheless, Jones and Lahey were pushing, insisting almost, that the sub they built for Vescovo be certified, primarily to add value for a potential buyer. "Certification of a full ocean depth submersible was a groundbreaking issue," Jones says. "Triton had never built a sub that had not been certified. Besides, many science organizations and other private individuals would never get in a sub if it is not certified."

Involving the certification company in the design phase was paramount, as the company would need to weigh in on decisions as they were made. Lahey had in mind using Det Norske Veritas-Germanischer Lloyd (DNV-GL), a German-Norwegian company that is the world's largest maritime classification society. The head of its submersible unit, Jonathan Struwe, thirty-four, was one of the most knowledgeable people in the world on submersibles, making him the ideal sounding board as the company built the sub.

"DNV-GL was far more receptive to the ideas being promoted for this new vehicle which was breaking from traditional configurations," Lahey said. "Jonathan became a co-designer from day one. He worked hand in glove with John Ramsay and added many of the clever design ideas."

Vescovo was eventually persuaded that a certified, two-person sub was the best idea. "It was clear that having a commercial vehicle that could execute tens of thousands of cycles to many depths over many decades was the way to go, and certification is what makes that possible," he said.

By June, the focus of the material for the hull had shifted from glass to metals. "That is a pretty fundamental decision: what are you going to make your house out of, brick or wood," Lahey says. "What to make the hull out of was the most significant initial decision that had to be made."

The issues with the hull were its physical diameter, the cost of material, and the corrosion properties. Furthermore, the diameter mattered because the hull could not be wider than the largest pressure-testing chamber in the world at the Krylov Shipbuilding Research Institute in Russia, or an entirely new facility would have to be built at an enormous cost.

Another variable was weight versus buoyancy. Regardless of what metal was used, the hull would weigh up to 8,000 pounds. A projection had to be made as to how much syntactic foam would be needed to make the sub buoyant on the surface and also neutrally buoyant underwater so that it would not sink into the ocean floor.

Syntactic foam is a material created by using hollow spheres made of glass that are bound together with a polymer resin. Though the foam itself is heavy, the ceramic spheres provide buoyancy. It is the only material that could create enough buoyancy for a multi-ton sub while also not being crushed by pressure.

The hull and the foam would have to withstand 16,000 pounds per square inch of pressure, the equivalent of 290 fully fueled 747s stacked on top of it. Several types of steel came under consideration, including duplex stainless, nickel-rich steel, and Inconel (an oxidation corrosion resistant material), as did titanium. Any metal was believed to be better than glass because metal is more forgiving under pressure. If metal were to fail as the depth increased, it would do so slowly and constructively, rather than suddenly and potentially catastrophically like glass. A metal sphere would also be easier to build from an engineering standpoint.

Ramsay began to explore the various metals. "I am not opposed to using titanium, steel, or possibly even aluminum for the pressure hull and realize that it all comes down to cost," Ramsay emailed Vescovo on June 2, 2015. "The reason I specifically mentioned titanium was because of its high strength to weight ratio, but that's not to say it would lead to the best value solution."

Still, titanium was the most attractive solution for several reasons. For starters, its properties are well known, and it is half as heavy and twice as strong as carbon steel. In addition to strength, titanium doesn't corrode in seawater. A further factor was that the price of titanium had dropped by two-thirds over the past ten years.

Another advantage was that a titanium hull could be built without any welds. Two large pieces could be machined and bolted together. The steady increase in pressure as the sub descended would seal them tighter. This was important because welding introduces discontinuities into the material that ultimately create localized stresses that could undermine its strength. The fact that Triton could build a spherical hull and precisely bolt it together to the required, incredibly tight tolerances rendered titanium the winner.

Triton began to delve into other major aspects of manufacturing. Two additional keys to the sub were finding high-voltage batteries that would power its systems and the syntactic foam that would be wrapped around the hull to provide buoyancy. The batteries would need to withstand the pressure and be able to deliver a substantial amount of power to the electrical system and the thrusters that would propel the sub. It was an estimated eight-hour round trip to the bottom of the Challenger Deep, not counting the time spent exploring the bottom.

Triton turned to Ron Allum, who seemed like an ideal collaborator. Based in Australia, Allum had designed and built James Cameron's *Deepsea Challenger.* He and Lahey had worked together on several projects, and Lahey considered him one of the best sub designers. He owned a company that manufactured batteries and syntactic foam.

Vescovo paid for Jones, Lahey, and Ramsay to fly to Sydney, Australia, for a five-day "sub summit" with Allum.

The meetings were productive, and the Triton team came away encouraged. Allum had insight on what needed to be improved on from the *Deepsea Challenger* to build a sub that could do hundreds or even thousands of dives over decades, rather than just one. His recommendation was to use Titanium Grade 23 for the hull. Though it would be more difficult to forge and machine than other grades, Grade 23 had a lower oxygen content than most grades, a quality Allum believed would make it more durable.

In the weeks after the summit, Allum started pressure-testing his batteries and syntactic foam, but it wasn't long before red flags popped up. The cost estimates for Allum's batteries and foam came in far higher than Triton wanted to spend. But Lahey said the real problem was that

Allum's lawyer insisted that he could not provide any warranties for the components, even if they failed because of defective workmanship. Lahey pleaded with Allum to provide the warranties, but he wouldn't budge.

"We were always a little concerned that he was a one-man show who didn't have much external support, and his financial position wasn't very good," Jones says. "The thing that really killed the deal for us, though, was that he wanted to charge roughly $1 million for batteries and $1 million for foam, and he wasn't willing to guarantee it at all. We couldn't risk our client's money without a warranty."

Jones called Vescovo and filled him in. "Ron is trying to build his retirement on you," Jones quipped.

Triton would have to find other suppliers for these two critical elements and proceed without Allum. Nevertheless, Vescovo felt confident enough with these initial findings to move forward with a comprehensive design plan and cost analysis.

❧

Vescovo had already formed a corporate entity named Caladan Oceanic LLC for the venture. He named the company after the ocean home world of the Atreides clan from the science fiction novel *Dune*, his favorite book, as he was now building his own unique ocean world, so to speak. So, the sub, the support vessel, and all financial and legal activities for the project would happen under the umbrella of Caladan Oceanic.

Vescovo brought two trusted advisers aboard Caladan, first attorney Matt Lipton as general counsel, and later a chief financial officer, Richard (Dick) DeShazo.

Lipton met Vescovo when they were classmates in the seventh grade at St. Mark's School of Texas. The two grew up near each other in a middle-class neighborhood in North Dallas and forged a lifelong friendship.

As an attorney, Lipton had counseled on several projects over the years for Vescovo's private equity firm, Insight Equity, and also for Vescovo personally. At the time he began working for Caladan, he was living on a vineyard two hours from Dallas with his future wife, trying to grow the winery and sell the brand.

"Victor is the type of person who will come to me and say, 'I'm buying a helicopter in Canada. Can you paper it, complete the transaction, and get it through customs to Dallas?'" Lipton says. "That type of thing is somewhat normal for him."

But Lipton wasn't quite sure what to make of Vescovo's plan to have a submarine built to dive to the bottom of the five oceans. "He did the seven summits and skied the poles to test himself, but other people have done that," Lipton says. "This was different. No one had done it. It was a challenge of adventure, technology, and business skills. I call him the most interesting man in the world, and I thought, if he can pull this off, then maybe he really is."

As the process ramped up, Vescovo hired Dick DeShazo to be CFO. The project was certain to be complex financially and also have its rough edges. DeShazo was brought on both for his financial acumen and his even-keeled demeanor.

A 66-year-old Southern gentleman with an accent to compliment his upbringing, DeShazo downplays his CPA credentials, such as the fact that he holds a master's degree from Birmingham-Southern College and has been chief accounting officer of a New York Stock Exchange company. He often quips that he's the product of a public school education, including his undergraduate degree from Auburn University.

DeShazo had been hired by Insight Equity in the summer of 2009 as CFO of Allied Energy Company, a fuel distribution company based in Birmingham, Alabama, near the town where DeShazo lived. The company went through some difficult times during the oil price crash of 2008–2009, but it was eventually turned around and rolled into Emerge Energy Services LP, which was taken public in May 2013.

DeShazo stayed on for two years, commuting to Dallas, and then retired in 2015. Two weeks later, Vescovo brought him back to run due diligence on a division of the company that was being sold to Sunoco. That lasted until August of 2016 when DeShazo retired again—until Vescovo called a few weeks later.

"Victor said, 'I have personal project, would you like to become involved?' I replied, 'I don't need to think about it, just sign me up,'" DeShazo recalls. "He is the smartest the guy I've ever worked with. He's got the latest chips running between his ears, I'll guarantee you that. But what people don't understand is that he is 'regular people' at heart."

ஓ

On August 10, 2015, Vescovo and Triton entered into a formal Design Services Agreement (DSA). Under the eight-page agreement, Triton agreed to perform the design tasks necessary to fully define the vehicle, provide a cost estimate for manufacturing, and a time frame that the company felt was realistic to engineer the world's first submersible commercially certified to full ocean depth. It would also begin to research possibilities for a ship to carry and deploy the sub, as well as planning an around-the-world expedition to the deepest points of the five oceans. After the completion of the plan, pegged at 180 days, Vescovo could elect to proceed or abandon. For this, he agreed to pay Triton $910,000.

"I was pretty sure they could do it, so this was really a down payment and 'earnest money' to show Triton I was real and to kick this off—and make sure the sub would come in at a somewhat reasonable cost," Vescovo says.

Lipton, who negotiated the agreement with Triton on Vescovo's behalf, found himself on uncertain ground. "Quite frankly, I'm not sure we knew the full extent of what we were getting into," he says. His words would be prophetic.

CHAPTER 4
LIMITING FACTORS

The search for a magical alchemy in the design of the full ocean depth submersible began in the head of John Ramsay. The 35-year-old principal design engineer for Triton Submarines was tasked with conceiving something that no one had ever done. For him, it represented the kind of challenge that he had always wanted, and one that did not intimidate.

As a kid growing up in the UK, Ramsay loved building models. He would spend hours in his room, drawing contraptions that could sail, fly, or traverse land. He followed his childhood passion to the University of Glasgow and the Glasgow School of Art to pursue a degree split between product design and mechanical engineering.

While earning the dual degree, he forged an understanding of how the aesthetics of design and the technical aspects of engineering coalesced in unique machines. However, he found that true visionaries in their fields had something of an X factor that could not be taught in a classroom.

"University just wasn't for me," he says. "I'm not sure that I found it all that relevant. To be honest, I'm not really sure you can teach design. Engineering, yes, but not design. By design, I mean a plan or drawing produced to show the look and function or workings of something."

When he entered the world of submarines, Ramsay found a field that had few design experts, and an industry that was, in many ways, a blank page. Unlike the modern automobile industry, where designers make generational refinements to a very well-defined package, sub builders were free to make their own way and to color outside the lines. Without any "industry standard" starting point, designers were free to make revolutionary leaps forward in the vehicle's configuration.

Ramsay, who began designing submersibles for Triton in 2006, doesn't regard a sub as an esoteric subject or a mystery that needs to be solved. Where others see the marvels of technology that allow a man to ride inside a bubble immune to water pressure, he sees something much more simple that allows him to start the design of every sub, regardless of its complexities, in the same place.

"A sub is a pressure-resistant shell of some sort with people, life support, and electronics on the inside, and a chassis, buoyancy systems, and thrusters on the outside," he says. "Nothing is that crazily complicated. It's just a difficult exercise of packaging. You start by defining what goes into the sub, then arrange everything to make it as comfortable, lightweight, and compact as possible in a package that meets all of the performance requirements. Then, you progressively commit to more and more aspects of the design as parts enter production and close in on the final design."

Still, the full ocean depth submersible commissioned by Vescovo, which would be aptly named the *Limiting Factor*, would be far different from the previous subs that Ramsay had designed. The primary concerns were that it be able to withstand the pressure at the bottom of the deepest ocean repeatedly, communicate with the surface, and go down and come up in a timely manner. Other requirements, such as being able to see outside, aesthetics, comfort, endurance on the bottom, underwater speed, and stability on the surface during recovery, were all secondary.

"On a normal 3,300/3, you have to make it comfortable for billionaire passengers who want to have a drink, eat sushi, and relax while marveling at the sea life," he says with utter sincerity. "You are thinking about optimizing their view, surface stability, and making the sub appealing to look at. With the *Limiting Factor*, it was about boom!—depth—getting there and getting back. It was so pure in its objective."

~

Holding all of the elements of the final machine in your head simultaneously is the gift of a virtuoso designer, and Ramsay was known for his ability to do just that. His process was a bit unorthodox. Before drawing anything, he formulated a vision of the submersible in his imagination as to how form would follow function and vice versa.

"The first job is to think and think, and then think some more about aspects you know will be problematic, and to build up a mental understanding of where you are going so that it is second nature to what you are trying to achieve," he explains. "It's not a conscious thing of making a list and checking things off. You start by building up this mental understanding of what you are going to have to do to make it work before you can even really start to develop the concepts."

The centerpiece of the sub, a titanium hull, would be negatively buoyant, as opposed to the acrylic-hulled subs that Ramsay had usually designed for Triton, which were positively buoyant. This meant that the sub would have to be much taller to accommodate enough syntactic foam for buoyancy, as the hull would weigh several tons. However, the drawback of a tall sub was it would rock on the surface, so that needed to be considered.

Pressure was driving virtually every design element. The hull and everything placed outside of it would have to withstand full ocean depth pressure of over 16,000 pounds per square inch, and be designed to distribute the stress equally at ambient depths. The extreme pressure dictated that the oxygen tanks would have to be inside the hull because even the strongest oxygen bottles would likely implode below 3,000 meters. The surfacing weights on the bottom of the sub needed to release electromagnetically. Cables on the outside operating those would need to be in oil-filled junction boxes to prevent corrosion.

Over the course of two months, a vision of the submersible bounced around in his head constantly, whether he was sitting in his office or walking through the supermarket. When Ramsay felt he was close, he sat down and sketched out a few drawings.

He sent the drawings, along with preliminary mathematical calculations, to the Triton manufacturing and operations team in Vero Beach. The two sides started going back and forth over the workability of the design and discussing modifications. After each conversation, Ramsay would alter the design.

"I spent weeks working on a design that would satisfy all the concerns," he says. "But I finally realized that there was something fundamentally wrong about the design, so I had to try and take a few steps back and try and reconsider everything."

His primary frustration was that every design had an overhang because the blocks of syntactic foam needed for buoyancy were larger than the hull, and thus overhanging at the bow, creating a feature like the peak of a baseball cap over the hull. The advantage of this arrangement was that the sub could move forward much faster. The drawback was that overhang restricted how close the sub could come to view underwater features because the viewports were beneath the overhang. The overhang was also acting like a large fin, restricting the vehicle's streamlined shape for vertical movement.

As he thought on the problem, Ramsay decided to make a compromise: *Viewing would be more important than forward speed.* He reasoned that if he backed off the concept of the sub traveling like a forward-moving train car through the water and twisted the image of the train car sideways, then he could produce clean views from all the viewports, which would be placed on one side rather than in the front. The two pilots would sit facing the viewports. This also allowed him to put the thrusters, the propellers that would move the sub through the water, on the front and back ends, rather than on the sides.

There would be ten thrusters in all. Four thrusters would be used for transverse maneuvering and rotation, essentially driving through the water, and two additional ones for slow-speed maneuvering in tight spaces. The four others would be vertical thrusters used to slow the sub's descent and for vertical control.

"After months of developing traditional frame concepts to protect the thrusters from entanglement risks, I had the idea to make the thrusters ejectable, something that, as far as I knew, had never been done before," he says.

This design breakthrough with the new configuration also resulted in other gains. Even though a tall sub would not be very streamlined as it drove forward through the water, it would have an elliptical profile that allowed for greater speed on both the descent and ascent. The "up" and "down" time was a key factor because the sub would be traveling as far as seven miles in each direction. Also, the arrangement had a relatively streamlined profile from the side, allowing the pilot to traverse significant distances during the dive.

Ramsay sketched the revision. It looked like a large pillow placed upright. The three viewports at the bottom made it appear as if the sub had a face with a high forehead above the eyes. When Patrick Lahey first saw it, he joked, "It has a face that only a mother could love."

The blocks of syntactic foam would be placed *around* the hull. The main ballast tanks would be on top. Before descent the tanks would be full of air to keep the sub buoyant; and affixed to the tanks would be electric pumps that would fill the ballast tanks with water to sink the sub. Between the ballast tanks was a tube through which the pilots would enter and exit the pressure hull. The ten thrusters, angled in different directions for upward and downward and forward and backward propulsion would go on the port and starboard sides of the sub. The batteries would be placed outside the pressure hull, tucked in between the large blocks of syntactic foam.

"I pretty much felt I had it," he says. "It all went back to the simplification concept. If you have a shape that is symmetrical side to side, front to back and top to bottom, then you can focus on designing one-eighth of the shape of the sub, because everything is mirrored. Having a symmetrical arrangement also makes manufacturing less complicated since you can use the same parts on both sides of the vehicle."

Ramsay designed the thrusters, batteries, and manipulator arm to be ejectable in the event the sub needed to make a fast emergency ascent, or if they became entangled, one of the most serious dangers to a submersible.

After Ramsay finished his concept of the design, the process shifted to what Triton needed to commit to in the near term and what could be pushed off until later. Triton began searching for suppliers for the key elements such as the titanium, the batteries, the syntactic foam, and certain hard-to-manufacture electrical components.

It was an exhaustive process that spanned the globe. The titanium was found in North Carolina. The batteries would be built in Barcelona with elements that came from South Korea. Trelleborg, a Swedish company with a manufacturing operation in the UK, was contracted to make the syntactic foam.

After six months of each department weighing in, Lahey pulled everything together and reached the point where he was confident that Triton could build the sub. He put together the multimillion-dollar cost estimate for the sub that he believed was within the upper range of what Vescovo indicated he would fund. The estimate included spares of everything—foam, batteries, thrusters, electrical circuitry—in case of a failure when the sub was on the ship at sea. It also included a $500,000 line item to commercially certify the submersible, as Triton would work hand-in-hand with the certifying agency DNV-GL and its inspector Jonathan Struwe to ensure that the necessary commercial safety guidelines were met at every step of the way.

Lahey, Bruce Jones, and Vescovo then hammered out an agreement whereby Triton would build the sub at cost, without any markup. The company would receive major bonuses for meeting certain benchmarks including commercial certification and completion of the expedition. If the sub were sold after the expedition and, depending on the price, Vescovo agreed to pay Triton above the raw manufacturing cost. Vescovo also agreed to contract Triton's personnel, at cost, to operate the launch and recovery of the sub during the expedition. He negotiated these terms hard with Triton because he believed he was putting up 100 percent of the funding, with no guarantee of success, and that Triton would gain significant brand and marketing value from a successful completion of the mission.

Vescovo also required that all parties to the project sign strict non-disclosure agreements (NDAs) regarding the project. If there were major setbacks, he didn't want himself or Triton to spend precious time or energy explaining to others what was going on. His own experience with extreme technology development had trained him that it was best to conduct bleeding-edge development in stealth mode. In that way, the team would not be held hostage to artificial deadlines or the scrutiny and negative comments from outsiders.

Under these terms, Vescovo green-lighted the construction of the *Limiting Factor*. He had plenty of experience in complex projects. For example, just one of the five companies he was chairman of at his private equity firm was an electronics service company that manufactured high-reliability semiconductors for the aerospace and defense industries. Another firm where he was chairman machined high-strength metals, and yet another built high-end defense munitions like laser-guided rockets and subhunting sonobuoys. But his time was limited and he was not an expert in submersible construction. He needed someone—not affiliated with Triton and an expert in subs—to serve as his eyes and ears as the submersible was built.

During the design phase, Lahey had introduced Vescovo to Paul Henry "P. H." Nargeolet, a legendary figure in the submersible world. In September 2016, Vescovo met with Nargeolet in New York and hired him on the spot. He was pleased with Nargeolet's input on the design, so he retained him for the construction period.

Born in France and living in the U.S., Nargeolet, who turned seventy in 2016, had done more than 150 sub dives, including thirty at the *Titanic* wreck. After serving for twenty-five years in the French Navy and rising to commander, he ran the French government's deep diving program and was in charge of the deep submersibles, the *Nautile* and *Cyana*. He developed a reputation as the guy to call for a tough deep diving mission, such as the 2009 search for the Air France flight that crashed in the Atlantic Ocean. Almost unbelievably, he had helped lead a team that found and retrieved the aircraft's black boxes on the very bottom of the ocean.

"I had been involved in some amazing projects, but for me this was truly unbelievable," Nargeolet says. "It would be a breakthrough for deep ocean diving."

⁂

The first step in construction of the *Limiting Factor* was to build its titanium hull, the sphere that the pilots would sit in. Vescovo and Ramsay had an extensive review of metals and mutually concluded that titanium would be the best bet, for workability, stability, and cost.

During a meeting in Hamburg, Ramsay met with Struwe and a metallurgist at DNV-GL to discuss the grade of titanium that should be used. They even reran calculations with some steel-based alternatives to make sure they had the strongest and lightest material for the hull.

They reaffirmed the strength-to-weight ratio advantage of titanium over steel. A titanium hull would weigh less, meaning less syntactic foam would be needed. Less foam would mean the submersible would be smaller and thus easier to store, launch, and recover on a support ship. The open question was what *grade* titanium to use—there are numerous grades, each with different characteristics, costs, and trade-offs.

Based on Ron Allum's recommendation, Triton had originally planned to use Titanium Grade 23 for the pressure hull material. But after taking advice from DNV-GL's pressure vessel experts and metallurgists, Ramsay revised the material specification to the more readily available Titanium Grade 5. He concluded that it was easier to work with and actually offered slightly better mechanical properties.

"Easier manufacturing of the hull is not the only benefit to titanium," Ramsay says. "Its relatively low density allows the hull to be exceptionally thick without being overly heavy. On externally loaded pressure vessels, thick walls are great. They are almost immune to buckling—the 'Coke can effect'—and, for our application, the additional thickness afforded by the titanium provides a great support for our eight-inch-thick viewports." All previous full-ocean depth submersibles had had just one viewport; at Triton's insistence in the pursuit of a scientifically useful craft, Vescovo had agreed to three.

Ramsay also knew from discussions with the Woods Hole Oceanographic Institution that the spheres of the *Alvin* submersible had issues with the welded joints of their pressure hull. After careful study and review with DNV-GL, the decision was made that there would also be no welds on the hull of the *Limiting Factor*. Instead, the hull would be held in place using an alignment ring and bolted brackets, with the ocean pressure itself providing the final sealing of the two halves of titanium. This also removed the need for post-welding heat treatment and eliminated any porous or other flaws in the welding that could fail over repeated dives to full depth and back. A lower-tech bolting of the two hemispheres turned out to be the best answer.

The viewports posed a major challenge. They would be made from acrylic, which is far more flexible and nowhere near as strong as titanium. In fact, the pressure placed on the acrylic would technically be higher than the rated strength of the material. Based on Ron Allum's viewports for Cameron's *Deepsea Challenger*, Ramsay created conical openings for the eight-inch-thick viewports with a slightly flexible seal around them so that when the sub descended and the pressure increased, the viewports would literally be pushed into the opening, thereby keeping them sealed.

Using titanium also allowed Ramsay to increase the size over Allum's design on *Deepsea Challenger*, which employed a single-person, quite small steel capsule, without any increase in the vehicle's weight. This was a directive of Jones and Lahey, who had convinced Vescovo that a two-person sub had far better onward sale potential than a single seater.

But the size of the hull was restricted. The first issue was that it had to fit inside a pressure testing chamber for certification purposes. The second was that the hull could not be more than 1,500 mm, or 59 inches in diameter. Export controls were placed on any vehicle above 1,500 mm, classifying them as military grade, meaning that a hull above that size would not be able to freely move from country to country. This meant that the two seats and all the necessary electronics and controls would need to fit inside a sphere less than five feet wide.

Ramsay addressed the tight space by angling the two seats slightly toward one another and putting the console with the joystick control in the middle. The oxygen bottles, allowing for a normal use of sixteen hours of oxygen with emergency supplies that could last ninety-six hours, were placed around the top of the capsule, along with carbon dioxide scrubbers under the two seats that would allow the occupants to breathe the same air over and over again.

"The life support system is obviously critical," Ramsay says. "It is designed to operate like a small, enclosed planet Earth, though without the plants!"

❧

When the design was finally locked in, an around-the-world production adventure began. The hull was forged in February 2018 at ATI Ladish in Milwaukee in a dramatic firestorm with titanium mined in North Carolina. The spheres were then shipped to STADCO in Los Angeles and then to Spain, where they were machined to within 99.933 percent of true spherical form and the holes for the viewports were carved out. The viewports were made in Germany and then and shipped to Barcelona for testing. Many electrical components for the communications system would be sourced from the UK or hand-built in Triton's facility in Florida.

To satisfy the DNV-GL safety requirements, all components of the submersible including those placed outside the hull had to be pressure tested to 1.2 times full ocean depth (20,000 psi). Working with the battery supplier, Triton developed not one, but two high-pressure testing chambers in Barcelona of different sizes for the components outside the sub, such as the batteries, cameras, thrusters, and electrical connections. Testing the hull was more challenging.

The pressure hull of *Deepsea Challenger*, though not commercially certified, had been tested at Penn State University, but that chamber was too small to hold the titanium hull of the *Limiting Factor*. The only pressure-testing chamber in the world that was large enough was located at the Krylov Shipbuilding Research Institute. Unfortunately, it was in St. Petersburg—Russia.

This posed significant challenges, of transportation, but also of politics. Given that the Russians weren't exactly trusting of the U.S., there was a concern that the Russian government might not believe this was a private enterprise, but rather a government or even military venture in disguise. At the time, the U.S. was also in the middle of numerous government investigations into Russian meddling in the 2016 presidential election. Overall, relations with Russia were not exactly warm. What if the Russians seized the sub capsule as some form of retribution, or even just to extract a hefty "exit tax" on the now-FOD-certified capsule?

It was a chance that had to be taken to obtain certification. The 8,000-pound hull was shipped to Triton's facility in Barcelona, where final assemblies of the viewports, penetrator plates, and hatch were done.

The hefty yet precious cargo was then loaded on an enclosed truck and driven 2,200 miles through five countries to the Krylov Institute.

Lahey and Struwe flew to Russia to supervise the testing and brought a translator with them. When they entered the Krylov facility, they were a bit unnerved. Rather than the high-tech sub testing lab that they had envisioned, the facility was a relic of the Cold War, a relatively run-down building with vines growing through its crumbling walls.

The hydrostatic testing chamber, named the DK-1000, was a massive underground tank. Before placing the pressure hull assembly in the chamber, the Russian crew connected a pipe to the sub's hatch and filled the hull with water to test for any leaks. This was done because if the hull were to fail in the pressure chamber, the shock wave would take out the entire building.

The two men watched nervously as the crane operator lifted the sub from its crate and maneuvered it into the testing chamber. As the hull was lowered into the chamber, it barely fit—to the point that it scraped the sides. Lahey briefly closed his eyes.

Once the hull was in the chamber, one of the techs connected a pipe into a fitting to fill the chamber with water. Lahey thought to himself, there is no way that fitting will hold with all that water blasting in there—yet it did.

The chamber then simulated an actual dive by gradually increasing the pressure on the hull for three hours on what would be the sub's descent, holding it constant for two hours to simulate the sub on the bottom, and then reducing it over the next three hours on what would be the sub's return to the surface. The process was repeated three times, taking a day and a half, but was completed fairly quickly so as not to draw too much attention to it.

Despite working on this somewhat mysterious U.S. project, the Russian testing team was rooting for success. When the gauges indicated that the hull passed the final time, the lead technician clenched his fist in the air and proclaimed: "Da!"

The pressure hull assembly performed like a rock, with no change in stress at any of the points. At the bottom depth, it was pressure-tested all the way to DNV-GL specifications of 14,000 meters, or 45,920 feet—roughly 1.2 times the depth of the Challenger Deep. The tests validated the design, as well as Triton's calculations and analysis of the entire pressure hull assembly. At a later date, the syntactic foam blocks that would attach to the hull would be tested the same way.

After the pressure hull assembly was removed from the chamber and drained, it was repacked for transport. It would be sent by air freight to Miami, and then trucked to Vero Beach. Lahey personally sealed the container in Russia and watched it leave the building.

A week later, Lahey apprehensively opened the crate the morning it arrived at Triton's warehouse.

"Man was I nervous when I opened it," Lahey says. "For all we know they could've moved that crate to a warehouse and shipped us a couple tons of rocks."

Thankfully, the pressure hull assembly was in the crate, intact, unharmed from the journey, and now fully tested to commercial safety standards.

Triton had built a mock-up of the hull out of fiberglass and foam to determine the ergonomics. It also built a frame out of wood to determine how the foam would attach to the hull. Using the fiberglass hull, the team set about figuring out how to assemble the electrical system and other interior components. They needed to determine the placement of every component from the seats to the electrical controls to the joystick that operated the sub. The builders would later lament that the sub sure looked great from the outside, but like the Lamborghini it was often compared to, it could be a devil to service.

The electronics were being designed and built by Tom Blades, Triton's principal electrical and systems design engineer. A soft-spoken, unassuming electronics whiz, Blades had come aboard Triton two years earlier. Both he and Ramsay worked out of the UK.

One of the challenges Blades faced was designing a communications system different than others he had done. The shallow diving subs had an ultra-short baseline system (USBL) to communicate with the surface that used acoustic pulses, but that system would not work in Hadal Zone depths. For the *Limiting Factor*, Blades would need to design a system for the sub to communicate with the ship using a modem that could transmit and receive analog communications as well as data. It would take seven seconds for sound to travel through the water column from the ship to the sub at full ocean depth. The modems Blades designed would have a time reference synchronized to the GPS on the ship, allowing Blades to calculate the speed of the sound through the water and thus determine the sub's depth.

The sub would need to communicate with the three landers dropped to the ocean's floor for navigational and scientific survey purposes. Landers are metal-framed, square scientific platforms with syntactic foam for buoyancy and weights to make them sink to the bottom. They are equipped with high-definition cameras to film the sub and sea life, bait traps to capture sea creatures, and depth and water salinity gauges. The landers required separate electrical systems to communicate with the ship's command center so they could be released from the bottom.

Blades also needed to design the control panels inside the sub that held gauges measuring depth and position, along with the monitoring and warning systems for oxygen and CO_2 levels. There would be two Toughbooks, the panels (or graphic user interfaces, GUIs, as they were usually called) used for data readout inside the sub, including the descent rate, the heading, the speed, and the depth. His approach was to duplicate each system and make them run on separate electrical circuits in the event that one of the sides, each of which contained three batteries, failed.

As the other components began arriving at the shop in Vero Beach, Ramsay was making constant refinements to the design based on functionality and practicality.

"Everything that came in was bigger than it was meant to be so you end up crowding the things into the sub," Ramsay recalls. "Tom and I used a virtual layout to study placement of the gear. We ended up putting much of it behind the seats."

The syntactic foam came in four large blocks weighing 1 ton each, with two additional blocks weighing 600 pounds each. Because the blocks were so large and heavy, how and in which order to attach the blocks to the hull became an issue. Their order determined how other components, such as the batteries and the surfacing weights, would be affixed.

Next came the positioning of the low pressure pumps that would pump out the water from the trunking that led to the hull's hatch when the sub surfaced. The final plans didn't allow enough room for them, as they were slightly larger than anticipated. More questions popped up. Where would the oil-filled electrical junction boxes be mounted? Where would the electrical wires be run? How would the batteries fit?

As the team began assembling the parts to the *Limiting Factor* in the spring of 2018, it felt like they were working on a toy model kit—without the instructions. Others likened it to a game of "submersible Tetris." "You can draw a picture and put a stickman in it, but the reality is much different," says Kelvin Magee, an experienced sub builder who runs Triton's machine shop. "We quickly figured out the sub couldn't be put together, or serviced on a ship, the way it was designed."

As Ramsay explains: "Ninety-eight percent of everything we produced was assembled, accessed, and serviced exactly as designed, which as you would hope, goes completely unnoticed. Of course, that leaves 2 percent of things that are not quite right, which then take up 98 percent of the time for the person who was to fix it—resulting in the entirely understandable misconception that nothing works!"

CHAPTER 5

CLASSING UP THE SHIP

The USNS *Indomitable* had a romantic history. Commissioned by the U.S. Navy in 1985, the ship was designed to hunt submarines during the Cold War. After its surveillance days ended, the ship was re-tasked to track down drug smuggling boats in the Caribbean. The ship was then transferred to NOAA, renamed *McArthur II*, and converted into an ocean research vessel. In 2014, a private marine company in Seattle purchased the ship and renamed her *Ocean Rover*. That was where Bruce Jones and Patrick Lahey found the ship in the spring of 2017 and targeted her, in an ironic twist, to deploy and support a sub rather than hunt them.

The Triton partners had encouraged Vescovo to finance a "turnkey" operation, which meant purchasing a ship to carry the *Limiting Factor* and its operating team around the world. They had cautioned against leasing a transport ship, as James Cameron had done with the *Mermaid Sapphire* during his 2012 expedition, because the expedition team wouldn't have full control of the vessel.

The expedition would span some 40,000 nautical miles and take a year, perhaps longer. In addition to the vast distances between the deepest spots of the five oceans, the expedition needed to separate by six months the trips to the two polar regions, the Arctic Ocean in the north and the Southern Ocean, to coincide with the summer in each region, thus ensuring the most favorable weather. Purchasing a vessel was deemed the best option, both for scheduling and for pairing a ship with the sub to create a system that would be more attractive to a buyer.

Vescovo agreed with their assessment, reasoning that leasing would make them beholden to the ship's owner and would also provide potentially crucial flexibility if something went wrong with sub, ship, or weather. "Deep down, I said how much more would I pay to have ultimate flexibility," he says. "Bruce and Patrick told me the ship was going to cost $5 million total, about the same to lease versus buy."

After Jones and Lahey viewed a number of ships online and a few in person, they felt that the *Ocean Rover* looked like the best option. Lahey called Rob McCallum, a longtime friend who runs the Seattle-based EYOS Expeditions, a company that plans and manages maritime and other expeditions around the globe, and asked him to take a firsthand look at the ship.

McCallum, who had worked on ship builds and refits, was serving as a sounding board for Lahey as Triton developed its relationship with Vescovo, though he was not yet part of the project. "When Patrick first called me, he said, 'I've found our unicorn,'" McCallum says.

McCallum was happy to help. He jumped in his car and made the short drive to the Lake Washington Ship Canal, where the ship was moored. He toured the vessel and came away with several concerns.

"The ship needed to be reclassed as a civilian ship because the standards for military vessels are lower, and that is expensive," he recalls. "Another concern was that the lifting gear wasn't heavy enough for the size of the sub they were building. The biggest concern was that it was in really bad shape overall. What I liked was that it was built strong and was quiet, but it needed a lot of work. I told Patrick not to buy it, or to buy it with his eyes wide open."

Jones and Lahey decided it was worth a closer look, so they flew to Seattle to see the ship.

Built in Tacoma, Washington, the ship is one of twelve stalwarts; its class was contracted by the U.S. Navy, six of which were built in Tacoma and six in Mississippi. The three-deck ship is 224 feet long, with 44 berths (expandable to 49), and carries 228,000 gallons of fuel. It is powered by four diesel generators that provide electrical power to drive the propellers, which have flexible rubber mounts to dampen the engine's vibration and make it a more effective sub-hunter.

Jones and Lahey liked many things about the ship. As a naval "hunting" vessel, it was exceedingly quiet, essential for underwater acoustic communications between the ship and the sub. The ship had both a dry lab and a wet lab. The dry lab could serve as mission control and house all the communications equipment. The wet lab would serve the science team as it recovered and examined marine life.

The ship had ample outdoor deck space that could be built out for operations. The stern had a large space where a climate-controlled hangar could be built to house the submersible. This was important not only for protecting the submersible from the elements during transit, but also to give the sub crew a covered area to work on the sub in extreme weather.

Though the engines had 100,000 hours on them, they were deemed in good condition. The ship was ice classed, meaning it was rated to travel through the iceberg fields of the Southern and Arctic Oceans, also a must.

But changes were needed. There were two cranes, a fixed A-frame crane on the rear of the ship and a hydraulic knuckle boom crane that swung out over the water. The A-frame would be used to launch the sub, while the knuckle boom crane would launch the tenders and the landers (the scientific platforms). The knuckle boom crane also needed to be able to recover the sub if the A-frame failed, meaning that it would need to be upgraded to hold a 12-ton sub. While that was doable, the A-frame posed a greater challenge. Because the A-frame was fixed in place, there was an issue with how far it extended beyond the stern over the water.

Jones voiced his concern of the crane's reach. "I remember quite clearly standing on the aft deck with Patrick and saying, 'This A-frame is not long enough. It needs to be extended and beefed up because you can't recover the submarine reliably in heavy weather this close to the stern. You are going to have problems with it banging into the ship.'"

Another drawback to the A-frame crane was that it was not "man rated." This meant that the sub pilots would need to board the submersible on or over open water. But neither Jones nor Leahy saw this as a problem, as most of Triton's subs were built this way, and John Ramsay had designed the full ocean depth sub accordingly.

Lahey did an extensive survey of the ship. Other issues seemed workable with refinements. The cabins needed updating, as did the galley and dining area. The bridge also needed several upgrades. All things considered, Jones estimated the refit and renovations would cost about $2.5 million—a number that McCallum warned Lahey privately was far too low. Jones, however, was confident he could make the aggressive number work because he himself would oversee the refit. He and his wife, Liz, were also planning to live on the ship and serve as expedition leaders.

"It seemed to us to be the ideal ship from the very beginning," Jones says. "We believed that if you didn't go crazy and start to rebuild everything and redo the interior that it was in good shape. I don't know that we looked at anything after that."

Once they had settled on the ship, they were introduced to Kyle Harris, a marine diesel engineer, through a client in the Pacific Northwest. The client recommended Harris because of his local knowledge and extensive experience. Lahey and Jones invited Harris to come on a sea trial on Lake Washington to observe the ship in action. The sea trial did not go well. Harris recalls several systems malfunctioned and alarms were constantly sounding.

"Let's just say that the ship needed a great deal of work to make it functional for an around the world expedition," Harris says.

Nevertheless, Jones and Lahey liked that it was ultra-quiet, fuel efficient, large enough to hold a support team, ice classed, equipped for science, and had a large mission deck. They both agreed that the positives outweighed the negatives and that it could be made ready for the expedition.

Based on their recommendation, Vescovo purchased the ship on March 14, 2017, for $2.5 million. The purchase of the ship turned out to be a decision that was not fully and properly vetted. The budget for the refit of the ship, set at just over $2.5 million by Jones, would fall far,

far short of what was needed. The decision not to upgrade to a man-rated A-frame crane would later be second-guessed to the point that it threatened to bring the entire expedition to a grinding halt.

"When you looked at it on paper, everyone agreed that it would work fine," Vescovo recalls. "We talked about a man-rated crane, but Triton said it would be $2 million and that we didn't need it. John Ramsay got involved and said, 'We'll use this A-frame to launch and recover, and we will enter the sub from the ocean.' When you are sitting in an office looking at it on paper, of course that works. But when you go out in reality with two-meter waves, it doesn't work as well. The problem is that Patrick and Bruce, despite all their experience, had never refit a ship before. And it is a very uncertain, black art."

⁂

With the submersible in the building stage and the ship purchased, Vescovo named the venture the Five Deeps Expedition, later adding the motto "In Profundo: Cognitio" (In the Deeps: Knowledge). He combed the pages of his beloved science fiction novels to name the submersible, the ship, and the support boats, and ended up turning to one of his favorite authors, Iain Banks, for inspiration. The worlds created by Banks in the Culture series had left an indelible impression on Vescovo, and he would use them for the real life expedition he was creating.

Vescovo also saw a nice parallel to Elon Musk's SpaceX venture. Musk, also a devotee of the Culture series, had named three of his rocket-recovery drone ships after spacecrafts from the Culture series: *Just Read the Instructions*, *Of Course I Still Love You*, and *A Shortfall of Gravitas*.

"As a tip of the hat to both the late Mr. Banks, as well as Mr. Musk, who is sending things *up* while we are sending things *down*, I decided to draw on the Culture series in naming the ships of the Five Deeps Expedition," Vescovo explains. "Science fiction propels forward thought, and makes curious people want to build the things that authors have dreamed up. Our effort to reach places no one else has reached would require engineering and imagination."

After considering the many options from Banks's series, he settled on the names that were the most fitting and self-explanatory to the tasks of the vessels of the Five Deeps.

The submersible was named the *Limiting Factor*, owing to the fact that its design focused on what the limiting factors of the oceans were, notably extreme depths and how to survive there. The ship would be the *Pressure Drop*, as it would transport the *Limiting Factor* and drop it into the extreme pressure of the oceans' depths.

The ship would carry four boats, each serving a different function. The fifteen-foot protector boat, which would be deployed each time the sub dived, would independently monitor the *Pressure Drop* and the *Limiting Factor*, triangulating communication during the dives. Its name would be *Learned Response*. A standard Zodiac boat would transport the sub pilots to and from the sub, as they would be boarding while the submersible was in the water and linked to the ship by a tether. In keeping with the "z" sound, it was christened *Xenophobe*.

The rigid-hull supply boat was named *Little Rascal* because it, well, *looked* like a little rascal and would be running to and fro with supplies. The rescue boat was named the *Livewire Problem*, because, Vescovo figured, it would likely only be used in the case of a major human error, or "livewire" (as opposed to software or hardware) problem in engineering-speak.

The three landers deployed for science and navigation were named for a major set of Culture characters—AI drones that are extremely intelligent and very symbiotic with the humans they interact with, and who also name themselves. Their full names were Fohristiwhirl Skaffen-Amtiskaw Handrahen Dran Easpyou, Sprant Flere-Imsaho Wu-Handrahen Xato Trabiti, and Uhana Closp, which were shortened for simplicity, thankfully, a crew member later remarked, to just: *Skaff*, *Flere*, and *Closp*.

The *Limiting Factor* being the expedition's primary vehicle, it carried the designation DSV, for Deep Submersible Vehicle. The *Pressure Drop* was given the designation Deep Submersible Support Vessel, or DSSV.

"Victor always loved the game *Dungeons & Dragons* and now he had something near his own version of that," his attorney and childhood friend Matt Lipton said. "How many people get to name the elements of an oceaneering expedition?"

❧

In no time, the budget Jones had drawn up for the refit to the newly christened *Pressure Drop* was tossed out the window, and the refit team began winging it. There were so many unaccounted-for items that costs began to escalate at a rate faster than even weekly revised budgeting could keep up with.

The main issue driving up costs was "reclassing" the vessel from a military to a civilian-owned ship. Because the vessel was built in the 1980s by the U.S. government, it did not have to adhere to any regulations set by SOLAS (Safety of Life at Sea Convention), but to reclass the vessel as a civilian research vessel, those requirements needed to be met. Triton had grossly underestimated the amount of changes that would be required to go from one standard to the higher one.

The list was long and would take far longer to complete than Jones had anticipated. The ship was not up to the fire code required on civilian-owned vessels. All of the doors needed to be changed to modern, fire-rated ones. The structural insulation on all three decks, which was non-fire retardant and flammable, had to be removed and new, fire-rated, structural insulation had to be installed—an undertaking that would take two months and a massive amount of labor.

The electrical penetrations—exposed electrical cables running through the bulkheads and decks—needed to be enclosed. Steel plates had to be placed around the cable bundles that ran from deck to deck in the stairways to prevent fire and smoke from moving between decks in the event of a fire. In total, more than fifty plates needed to be fabricated and installed, with each one taking more than 200 man hours to complete.

Safety issues also had to be addressed. The ship needed new life rafts as well as their launching cradles and stowing brackets. All of the safety gear and lifesaving and firefighting equipment had to be purchased, tested, and certified. The motorized rescue boat and its launching davit were not SOLAS-approved, so a new boat needed to be purchased, which alone cost $300,000. The communications system was also antiquated and its satellite dish, network system, and virtually all other communications gear needed to be replaced or at the very least upgraded.

"It became apparent really quickly that the refit was going to cost millions more than expected," Harris says. "Once the sale went through, we realized the amount of due diligence that had not been done."

Vescovo was, as one would expect, not at all pleased. He blamed Jones and Lahey's inexperience with ship refits and lack of communication about the potential pitfalls. McCallum was in "I told you so" mode. But at the end of the day, they were stuck with a ship that needed a lot of work.

Further complicating matters and driving up costs was that the *Pressure Drop* was being reflagged from the U.S. to the Marshall Islands. The Marshall Islands has the third-largest ship registry in the world because of its favorable regulations. Flying under the Marshall Islands flag comes with all of the protections of a U.S. ship without the legal requirements regarding international crews and wages. However, reflagging the ship meant that every aspect of it would need to be reinspected and recertified by the Marshall Islands authorities.

Part of the problem was that Jones was supervising the refit from his home in Idaho, rather than on-site in Seattle, and communicating remotely with Harris, who was handling the day-to-day oversight. Jones had never undertaken a ship refit like this and was admittedly out of his depth. To make matters worse, he underwent a four-level spinal fusion during the refit. Because he was hanging on to the project as a career capper, he pushed through and attempted to continue supervising the work during what turned out to be a longer-than-expected recovery period.

During this time, Vescovo's company, Caladan Oceanic, was hit with a $1 million workman's compensation claim. A worker fell into a hole in the deck that he claimed he had previously reported, though there were no witnesses to his injury and—in Vescovo's view—the confined nature of the area made it extremely unlikely someone could "accidentally" fall there. Vescovo blamed Jones for not paying close enough attention to the project, and not arranging for proper insurance to cover the cost of the legal defense and any claims. In fact, there was no insurance to cover this incident. Jones countered that he expected Caladan to have purchased insurance as it owned the ship and was paying the workers.

With the bills piling up and the refit running behind, Vescovo began looking for someone to replace Jones. Vescovo was put in touch with a former Navy SEAL from Florida who had done several ship refits. He hired the SEAL on a day rate and flew him to Seattle to assess the project. Lipton and Dick DeShazo, the Caladan CFO, also flew in for the meeting. Lahey met the group at the ship, along with Kyle Harris.

"It was a shit show in a good way because all the shit came out," Vescovo says.

From the ship, Lipton called Vescovo during the meeting and told him the situation was much worse than expected. He also said that it would head down a rabbit hole if the Navy SEAL were hired. Not only was the cost estimate the SEAL gave more than ten times the original budget, the man's arrogance rubbed everyone the wrong way. "This guy is a bigmouth," Lipton told Vescovo.

Vescovo needed a solution, not a further problem. The expedition would require the group to work closely together, and he was being careful to choose his team accordingly. Vescovo told Lipton to end the meeting immediately and send everybody home. He needed to think through what to do.

"I needed someone I trusted to do oversight so I put Matt in charge of general oversight, which also turned out to be less than ideal because as much as I know he wanted to help, he was understandably in a bit over his head because he had never done a ship refit either," Vescovo says. "The big point is that Bruce and Patrick made the ship refit sound very perfunctory and easy because they were hanging their hats on the fact that it was class rated. But it was class rated to *military* specifications but not modern commercial standards of the Marshall Islands flag. When there is a change of ownership, you shift to the latest class ratings. Bruce and Patrick didn't anticipate that. A big mistake that could—and should—have been anticipated."

By this point, Jones and Vescovo were on the outs. Jones was unhappy with the way things were going on the project, and Vescovo had lost confidence in Jones.

Jones acknowledges the class issues but claims that Vescovo's desire to have a more comfortable ship led to increased costs. "The refit started to grow," Jones says. "What we originally agreed on started to change, and that's not an uncommon thing with Victor. There were things people were concerned about that I wasn't. Looking over our shoulder all the time were Dick DeShazo and Matt Lipton. They are not marine guys. They don't know anything about ships. I think they were concerned that my attitude was a little cavalier."

By default, Harris ended up in charge of the refit, with Vescovo personally taking on the role of oversight, because after all, it was his money. As his girlfriend Monika said to him when he groused about the problem, "There's an old Albanian saying: 'Only the master can get his own donkey out of the mud.'"

Though Vescovo had searched for a less expensive alternative, Harris was the most practical choice, as he was already on the job. Harris had never run a full refit project, but Vescovo was impressed by his attention to detail, work ethic, and directness. Vescovo wanted the bad news immediately and the truth always, and Harris was the type of straight shooter who delivered just that. At this point, Vescovo reasoned, those qualities were actually more important than alleged "experience."

Despite the original plan to move the ship to Florida where labor costs were cheaper, the refit was forced to continue in Seattle. The ship was literally stuck in port until both SOLAS and the Marshall Islands' inspectors signed off on all the changes and provided proper documentation to satisfy the American Bureau of Shipping (ABS), the classification society for maritime safety that had a reputation for being extreme sticklers for regulatory compliance.

By this point, there was no budget. Harris was working item by item.

"Triton's budget proposal was not a realistic figure for the project, but they were fairly set in their ways that they had come to the right choice at the right price point," Harris says.

He would round up multiple bids for each item and then discuss them with DeShazo. The two talked three or four times a day. Once a vendor was chosen, DeShazo would wire payment the same day, as the only way to build credit with vendors was to pay everyone immediately.

The ship was finally certified as seaworthy in early February of 2018, and the Marshall Islands flag was raised. On the stern, where a ship displays her home port, the *Pressure Drop*'s stern displayed its home port as Majuro, M.I.—ironic as the ship had never been there. A quickly cobbled together crew sailed the ship down the west coast of the U.S. through the Panama Canal and up to Houma, Louisiana, a Cajun port town where the work could be continued at a far lower price than in Seattle.

The crew, which included Jones's son Sterling, had been pieced together by Jones from referrals. Days before arriving in Houma, the crew got wind that they would be dismissed when the ship arrived, as a new, full-time expedition crew was being assembled. This shouldn't have been a surprise, as they were not guaranteed work beyond the transit. However, they did not take the news like professionals.

Harris, who had flown to Houma, boarded the ship the morning that the transit crew departed. The vessel was trashed. There was garbage everywhere, broken bottles on the upper deck, toilets overflowing, and rotting food in the trash cans and freezer, which had been unplugged. Without the new crew on-site, Harris hired the shipyard workers to clean and sanitize the ship so that he could resume work on the refit, yet one more unwanted cost.

Over the next four months, the extensive refit continued in Houma under Harris's direction, who had relocated to the Bayou town to oversee the work. The sewage tanks, fuel tanks, and water tanks were all pumped out and cleaned so that the vessel could go into dry dock, the process of removing it from the water. This allowed the exterior to be sandblasted, primed, and painted, and the propellers to be repaired.

An entire new control system was installed and the bridge was updated. Originally, the ship was only able to be steered from one position, in the middle of the bridge. New steering stations were added on the wings of the bridge and in the rear. This would allow the captain to see the water from all angles when launching and recovering the sub. Miles, literally miles, of new electrical cable were laid for a new control system.

During the refit, Vescovo flew his jet down to Houma to check on the status. As a former navy commander that had spent many months at sea himself, he was shocked by the poor condition of many parts of the vessel and the overall quality of the lodgings and mess for the crew. He gave Harris the green light to the make the ship livable. After all, the expedition team would spend several weeks at a time, and sometimes as long as a month, on board.

The ship has three main decks and was expanded during the refit to forty-nine berths. The upper deck houses single cabins with private bathrooms used by senior crew, the owner, and the expedition leader, as well as the captain's office. Moving toward the back of the ship outside of the main deck, there is a boat deck, housing the tenders, storage lockers, lifeboats, and emergency equipment.

Below the main deck is the forecastle deck. At the front are several cabins on either side of the hallway. Each cabin has a naval-style metal bunkbed, two metal wardrobe lockers, a metal dresser, a desk and a sink with medicine cabinet. Flat-screen TVs were added to the wall at the foot of each bed with an extensive menu of movies. Small portal windows provide a view of the water. A combined shower and toilet area is shared with the adjacent two-person cabin. In the rear of the forecastle deck is the expedition office, science office, and a large dry lab used for all the communications monitors. In back of that area is the wet lab with sinks and refrigerators for scientific research. The door at the rear leads out to the bi-level aft deck that holds the submersible and the cranes.

The main deck, a flight of stairs down from the inside of the forecastle deck, has additional two-person cabins with connecting bathrooms, as well as one four-person cabin, the kitchen and dining area, food storage lockers and freezers, a TV lounge, the hospital, and a small gym. The entrance to the engine room located in the bowels of the ship is also on this level.

The bridge containing all the ship's navigational equipment is above the main deck. On the top of the bridge, accessible by an outdoor staircase, is an observation deck that would be dubbed the "sky bar."

During the refit, all of the cabins received new flooring and sinks. An expanded reverse osmosis water system was installed, as the existing water tank of 17,000 liters (4,490 gallons) was insufficient to keep up with anticipated consumption on a fully manned ship. The counters in the mess area were replaced and the galley was completely rebuilt to increase the number and comfort of the seats. And finally, a high-end coffee maker that produced espresso and lattes was purchased and installed. Even though Vescovo himself didn't drink coffee, he knew from his navy time that on any ship, high-quality coffee was essential for morale, if not actual effectiveness.

❧

In early June, as the *Pressure Drop* refit continued and the departure deadline approached, tensions ran high in the shipyard in Houma. Harris and the new relief captain, Mike White, who would be part of the expedition, were at odds. Part of the conflict stemmed from the fact that both felt they had control of the ship, Harris because he was running the refit and White because he was the captain.

Things came to a head on June 6 when the *Pressure Drop* left Houma for Fort Pierce, Florida, where the final work, including adding the communications system, would be completed and the sub placed on board. As the ship sailed out of the winding canals of the Louisiana Bayou, it scraped the bottom. The mud and silt clogged the ship's "sea chest," the intake reservoir that draws water through it, and the ship ran aground. Without proper water intake, the cooling system of the engines shut down automatically and most power was lost.

"We had a local pilot, and between him and the captain the two of them decided not to follow the vessel in front of them, which was also a deep draft vessel, and go their own way," Harris says.

A tugboat had to be hired to pull the ship out of the mud. After an eighteen-hour delay, the ship was finally on its way to Fort Pierce. In the grand scheme of the rush to mount the expedition, it was more of a minor headache than a migraine. Vescovo had been so absorbed with the refit of the *Pressure Drop*, he was only just beginning to appreciate the darkening storm clouds appearing over the submersible's production schedule. By this time, in nearby Vero Beach at Triton's facility, the bigger issue was that the struggle to finish the *Limiting Factor* had begun to resemble a Sisyphean task.

CHAPTER 6
TEAM LEADERS

Bruce Jones viewed the building of the *Limiting Factor* and stewarding the Five Deeps Expedition as the swan song of his maritime career. He had been waiting for the opportunity to be part of a world's first in the submersible game, both for the achievement and the notoriety. Moreover, his family would also be a part of it, as his son, Sterling, planned to be on the ship as part of the Triton crew, and he and his wife, Liz, would serve as expedition leaders. The real capstone for him, however, would be using one of Triton's contractual discretionary dives to dive with his wife in the *Limiting Factor* to the bottom of the Challenger Deep, a place only three men had gone, making her the first woman.

But this was not Jones's party, so to speak, it was Vescovo's. In its submersible construction contract with Vescovo, Triton initially had two discretionary dives at the Challenger Deep, but Vescovo took one away as part of a business transaction. Experiencing a cash shortfall, Triton asked Vescovo for a short-term, immediate $1 million loan to bail them out of a deal with billionaire Ray Dalio, who had done several projects with Triton. Vescovo agreed to lend the money, in part because the last thing he needed was for Triton to be financially unstable.

As part of the terms, Vescovo reduced Triton to one discretionary dive. Jones felt it was punitive; Vescovo called it part of a business negotiation that allowed him to reallocate the costly dive for another—likely scientific—purpose. Either way, Triton, which repaid the loan in full with interest, was now down to one dive. Unbeknownst to Vescovo, however, Jones was still hanging on to the notion that he and his wife would dive the Challenger Deep. It was, after all, in the contract that Triton would get a dive.

In actuality, Vescovo was tiring of having to deal with Jones on any front. He attributed the millions of dollars of cost overruns on the ship's refit to Jones's lack of experience. He also had let go the ship's captain that Jones had recommended in favor of someone with more direct submersible launching experience. Most of all, however, Vescovo evaluated businessmen based on their ability to manage deadlines and costs well, and Jones had done neither. It didn't help that personality-wise they were very different: Jones was a person who enjoys the good life while Vescovo was, as Lahey described him once, "a Vulcan." By December 2017, it was mutually agreed that Jones would step aside as expedition leader in favor of an experienced professional in the field.

"With Bruce, it just seemed to me that he was way out of his areas of expertise, and seemed to care more about enjoying the expedition—almost like an intense cruise—and becoming famous for the expedition, than the hellish daily grind it would take to make it happen successfully," Vescovo said. "It was quickly becoming clear that Bruce had given me really inaccurate projections, was failing at what he was responsible for, and I'd made my living off quickly changing out leadership that wasn't working out. So . . ."

However, Vescovo agreed that Jones would continue to lead the formulation of a marketing plan to find an onward buyer for the submersible and the ship since marketing seemed to be his real passion. Triton was heavily incentivized to sell the sub, as the company could possibly receive a significant, formula-based commission based on the total system sale price—if it sold. But even here, Vescovo was highly skeptical of Jones's strategy. "He said, 'We'll get a bunch of billionaires together on the ship for beers and sell it to the highest bidder,'" Vescovo recalls. "Like, good luck with that strategy . . ."

"Look," Jones says, "we were all very bullish about the future sale of the system because it is so unique."

Patrick Lahey, Triton's co-owner, was well aware of the tension. He was caught in a tricky position between his partner of ten years and his most important client, all while overseeing the team that was building the sub.

"I'm the peacemaker," Lahey says. "That's a pretty common role I play with my partner. Bruce is a very opinionated, strong-willed person. Sometimes that ego can be in direct conflict with our clients, who also have pretty big egos. He and I have always had a stormy relationship. For me, the situation was deeply frustrating and disappointing on many levels. I'm from the school where the client is always right. Our customers—and Victor is no exception—have very high expectations, and I've learned what it takes for us to achieve those."

In this case, it was agreeing to sideline his partner from Triton's biggest project.

To lead the expedition, Lahey suggested Rob McCallum, a veteran expedition leader who had been serving as a confidential sounding board for Lahey for two years—and who had recommended against purchasing the ship for the exact reasons that were now causing the rapidly escalating costs. In late December, McCallum flew to Dallas to make his pitch to be expedition leader and to deliver him some costly news.

Rob McCallum is nothing if not experienced in dealing with high net worth individuals seeking adventure. In 2003, after a 25-year career with the National Park Service in his native New Zealand, McCallum worked as a technical adviser for the United Nations in Papua New Guinea. McCallum then started a new career in commercial maritime expeditions. He had been part of the *Mir*'s dives to the *Titanic*, and he had worked extensively with James Cameron on his ocean expeditions and as the coordinator of the test dives of the *Deepsea Challenger.*

At the time he met Vescovo, McCallum's company was operating sixty expeditions a year, including twenty-five to the Antarctic, mostly for billionaires who are tired of normal family excursions and want a legacy experience. In all, the firm he cofounded, EYOS Expeditions, had run some 1,200 expeditions.

"Our motto is that if it doesn't break the laws of physics, then it's only a matter of time, money, and another bottle of red wine," he says. "If you can imagine these things, then there must be a way."

McCallum was impressed with Vescovo's commitment to dive the five oceans. Serving as the expedition leader would only solidify EYOS's reputation as the go-to company for a complicated expedition. But McCallum had to walk a fine line; he had to explain the cracks he saw in the current plan while also trying to establish a relationship with Vescovo. He was wary of coming on too strong and too critical of Triton. However, he needed to be upfront, not only to gain Vescovo's trust, but also for his own sake if he were hired as expedition leader.

"I knew things were not going well, and I don't think that Victor knew about how badly they were going," McCallum says. "I was trying to get across to him that I am not a bad news bear. I am someone who is always truthful and here is the truth."

McCallum laid out the problems one by one. The operational budget that Jones had drawn up needed to be doubled. For openers, it did not include fuel for the ship or travel for the Triton team that would run sub operations. He told Vescovo that the ship's crew needed to be replaced again with a more professional one. The plan to have one cook on board and to expect everyone to clean their own cabins and rotate doing the laundry was not workable. Finally, he said, the proposed itinerary was simply not operationally possible.

Vescovo listened and absorbed what McCallum was saying. The meeting at a sandwich shop in Dallas lasted less than an hour. At the conclusion, Vescovo asked him to prepare a new budget and itinerary, and to recommend a new captain.

"Victor is a careful analyst," McCallum says. "He is extremely clever. And so by showing him where things were going wrong, he was quick to understand the problems and also trust me to tell him the truth."

McCallum prepared a detailed document of all the things that he felt needed to be addressed, both with the ship and the schedule. He recommended Scotsman Stuart Buckle, who had actually been James Cameron's captain during his 2012 Challenger Deep expedition. He also recommended Vescovo partner in some way on the science component of the expedition with Oliver Steeds of Nekton, a nonprofit ocean research organization that had worked with Triton's subs.

A month later, Vescovo and McCallum met again. Satisfied with McCallum's new projections and his overall philosophy, Vescovo signed a contract with EYOS to lead the expedition.

McCallum made the introduction to Buckle. Before the ship's transit to Houma, Buckle flew to Seattle to see the ship—about which he had his doubts—and to meet with Vescovo. An easygoing Scot with red hair, Buckle came to the project with the ideal credentials. Having worked in the ocean and gas industry, he had fifteen years of experience deploying vessels from ships in the world's most remote regions.

Though the pay was less than what Buckle could earn on an oil and gas boat, working for a private owner had its advantages. He was allowed to handpick his crew, and he would only need to captain the ship on the dives, as a transit captain would sail the ship from ocean to ocean. Like McCallum, he was concerned about the ship, though McCallum had cautioned him not to get into a discussion about it with Vescovo.

"When Rob called, he told me, 'I have a job for you that you are not going to be able to say "no" to.'" Buckle recalls. "However, he cautioned me not to be too negative about the ship because it was too late to replace it. It wasn't an ideal vessel in terms of its age, fitness, power, and ability to deploy the sub, and it was in terribly bad shape. But I felt that with the right people in place we could get the job done."

Less than five minutes into the interview, Vescovo decided to hire the master mariner. "It was immediately clear to me that he was one of those guys that was very, very good at what he did and was also a solid leader," Vescovo later said.

⁂

Among McCallum's other concerns that he expressed to Vescovo were that the film production team Vescovo had hired were positioned to have too great a role in the expedition. To have a detailed and accurate video history of the expedition, Vescovo had signed a contract with Emmy and BAFTA award–winning documentary producer Anthony Geffen, who runs the London-based Atlantic Productions, one of the most prolific producers of nonfiction content in the world. Vescovo agreed to grant media exclusivity to Atlantic to secure a broadcaster for a series on the expedition. His primary goal was to ensure that there was a first-rate program made to document the expedition. With all the refit overruns, Vescovo also didn't want to add even more costs merely to film the expedition. Finally, he thought, someone else was willing to write some big checks.

Vescovo had met Geffen through Jones in early 2017. Vescovo had heard from Jones about David Attenborough's Great Barrier Reef documentary made by Atlantic and the Triton sub that was used. He was also impressed that Geffen had produced several landmark nonfiction television productions, particularly in natural history and exploration, including a series of films with Attenborough. Geffen had a reputation for delivering first-class films and then pushing the envelope to promote them—with great success, as evidenced by a trophy case of awards from around the world in Atlantic's London office. Such a film or series could also help promote the sale of the submersible.

Atlantic Productions signed a contract with Vescovo's Caladan Oceanic in May 2017. Under the agreement, Atlantic would pay for the expenses of the film crew, with the exception of meals and accommodations on the ship, plus take responsibility for securing all funding for the development, production, postproduction, and distribution of the documentary or series. In return for Vescovo providing the platform for the series, Geffen agreed to split any back-end profits.

Geffen traveled to Dallas to conduct some initial interviews with Vescovo to build a reel that might interest broadcasters enough to fund the filming. The British filmmaker and the Texas investor found they had something unusual in common: they had both climbed Mount Everest. Vescovo had done it as part of his "seven summits" quest. Geffen had climbed it while making a documentary titled *The Wildest Dream: Conquest of Everest*, about Conrad Anker, the climber who discovered the body of climber George Mallory, who had died attempting to summit the mountain.

Back in London, the reel Geffen and his team produced evidently worked. Talks for a series on the expedition were productive, and he was able to secure multiple meetings with interested parties.

Meanwhile, the schedule was coming together. The first dive of the expedition was scheduled for early September 2018 at the Molloy Deep in the Arctic Ocean. The ship would then sail back to Puerto Rico for the second deep, the Atlantic Ocean. On the way there and back, it would pass over the *Titanic* wreck, which was off the coast of Newfoundland. No manned submersible had been to the *Titanic* since 2003, so why not dive it?

Plans were made to bring aboard *Titanic* experts and film the wreck with high-definition cameras to evaluate its deterioration. Geffen would make a separate *Titanic* film and sell it as a special.

Geffen typically presold his ideas to broadcasters so that they were involved in the editorial process from the get-go. But this project proved to be a challenge for the standard funding model. While there was interest in the story, it was proving next to impossible to actually sign on a broadcaster and secure cash funding at this early stage. The *Limiting Factor* was not yet operational, so there was no guarantee the sub could actually reach the bottom of the five oceans. This meant that Geffen would need to invest Atlantic's money to film a couple test dives that showed the sub in action. Though it was a risk, Geffen had confidence that Triton could deliver a working sub. Some of his confidantes weren't so sure.

"I kept thinking, how is this man going to achieve this and what does it mean if he does," Geffen recalls. "The more I thought about it, the more I saw it as a story of human ambition meeting technical challenge, adding to the fact that this would happen in the deepest ocean waters where no one had gone."

Atlantic had filmed in some of the most remote places on the planet in conditions that ranged from the heat and humidity in rainforests to arctic temperatures. But filming at full ocean depth posed a new challenge. "The big issue was how to get good clear footage from outside of the sub," says Ian Syder, Atlantic's head of production. "There is a level of detail when you are talking about deep sea operations which is often beyond what we do in television. Atlantic had used underwater cameras on other series, but these cameras would need to function in much deeper waters."

McCallum, however, was concerned that having a film crew on the ship at all would interfere with the core mission of diving the five oceans.

He also felt that the TV company having control of all the media was not in the best interest of the expedition.

"What was not going to work was to have the whole thing led by a film production company because they are only interested in one thing—making the film," McCallum says. "I'm the Antichrist of film production because they are wanting unscripted disasters and fuckups, and I'm wanting control and calm." The drama and "reality" that made for good television was antithetical to what McCallum was trying to achieve.

Jones and Lahey didn't like the media deal Vescovo made because they felt it limited what Triton could do to promote the sub. Because Atlantic initially, and the broadcaster ultimately, would have de facto control of the media during the expeditions, Triton would effectively have to ask permission to release any information about the dives. "Victor made a really bum deal," Jones says.

Geffen, however, points out that it was the only way any broadcaster would sign on to the project. "What Triton failed to understand is that there are *a lot* of diverse interests involved in any expedition of this kind and caliber—not just Triton's own interest in promoting its brand," Geffen explains. "Additionally, Triton didn't understand that broadcasters require media exclusivity so as to justify their investment in filming the expedition and producing the film."

The idea was supposed to be that Atlantic (along with any broadcasters it brought to the project), Caladan, EYOS, and Triton would work together to mutually agree on press announcements relating to dives and the discoveries. But EYOS and Triton were never fully on board with the strategy, which would cause significant friction.

∽

To lead the science piece of the expedition, Lahey suggested Dr. Alan Jamieson, a researcher and senior lecturer at Newcastle University in Scotland. Now that the submersible would have viewing portals and a hydraulic arm to collect samples, the expedition was looking to add a significant science dimension. Conducting science was also important to

Geffen, as it added an element of interest for broadcasters and ultimately viewers beyond a man doing "hero dives" to the bottom of the oceans.

"When I first approached broadcasters with the idea, one of them said, 'I'm not going to fund some billionaire's gap year,'" Geffen says, punctuating the thought with a chuckle. "There was a real resistance to seeing past the idea that some rich guy was building a toy to break a record. But when Victor committed to do scientific exploration, the interest level increased. But that interest was always subject to the sub actually working. That was the unknown."

Jamieson was one of world's leaders in the exploration of the Hadal Zone and the ocean's subduction trenches. The 42-year-old scientist had participated in more than 50 deep sea expeditions and supervised more than 250 lander deployments. He had *literally* "written the book" on the science of the very deep ocean—among his published works was the book *The Hadal Zone: Life in the Deepest Oceans.*

Vescovo met Jamieson in London over lunch and liked both his credentials and his very direct, plain-speaking personality. An affable Scot with a thick accent and dour wit, Jamieson debunked the myth that the pilots of the *Trieste*, Jacques Piccard and Lt. Don Walsh, had seen a vertebrate on their 1960 descent to the Challenger Deep. "Their portal was tiny and there was ocean dust everywhere," Jamieson told Vescovo. "I'm telling you that you are not going to find a critter down there, but if you do, I want it! That will be a big deal."

To further entice Jamieson, Vescovo offered him exclusive use of the ship for a week to conduct scientific research in the Puerto Rico Trench in the days leading up to the deep dive there.

The two also discussed the fact that no one had definitively identified the deepest point of the Indian Ocean. It was believed to be in the Java Trench off the coast of Indonesia, but Jamieson indicated there was evidence that the Diamantina Trench near southwestern Australia could be home to it. Because Vescovo was looking at installing a massive multibeam sonar to collect data and have a hydrographer map the trenches, the expedition would be able to answer that question with certainty.

Vescovo contacted Newcastle University to bring Jamieson aboard. Vescovo reasoned that the university should just "loan" him to the expedition since the institution stood to gain from Jamieson being on a potentially historic mission, but university officials asked him for over $500,000 in compensation. Vescovo laughed at the ridiculous ask and began to look for alternatives at the oceanographic institutes of Stanford, USC, and UCLA. Weeks later, the Newcastle administrators came back to him with a modest five-figure number that he agreed to pay.

Jamieson was excited at the possibilities. He planned to bring fellow scientists on board on a rotating basis, including Johanna Weston, a Newcastle PhD student focusing on the Hadal Zone, and Heather Stewart, a marine geologist at the British Geological Survey who had done extensive work on the processes shaping the seafloor.

There were several potentially groundbreaking science opportunities that were all intertwined. The sonar would allow the scientists to examine the whole of each trench. The sub cameras would record what the seafloor in each ocean actually looked like as it traversed the bottom. The cameras would also capture which sessile (nonmobile or slow-moving) animals lived at the deepest points of the Earth.

The *Pressure Drop* would carry three landers, which Jamieson would help Triton design. He planned to bring along two more of his own landers. The lander cameras would record the larger, mobile animals to see which ones were present and which ones absent in each ocean, as well as searching for new species and researching population densities. The lander traps would capture the smaller animals, allowing the scientists to study their genetic adaptation to high pressure and phylogenetic connectivity, or evolutionary diversity, between depths and between locations.

Jamieson's plan was to gather core samples of sediment and rocks from all five oceans and conduct a microbial study on them. He would then be able to determine if they were all the same, all different, or one or more were the same. This could help in writing the story of the interconnectivity of the oceans.

"It was essentially all about biodiversity across the deepest 50 percent of the oceans, and by doing it across as many depths and locations as possible, we could then disentangled things that are true on a local scale and things that are true on a global scale," Jamieson explains. "The 'bonus balls' are always new species, plus 'money shots' of rare species and hopefully even the occasional downright weird shit that the deep sea sometimes throws up."

⁂

With the team rounded out, McCallum set up a two-day meeting in April 2018 of the principals of the expedition in Dallas. At that time, neither the ship nor the sub were ready for prime time. At all. The ship was still undergoing its refit, and the sub was being built section by section in Triton's warehouse and awaiting delivery from various suppliers—many of them late—before it could be fully assembled. The expedition schedule called for sea trials of the sub in open water to be held in early July in the Bahamas so that the ship could make the early September summer weather window at the Molloy Deep in the Arctic Ocean.

Jamieson, Buckle, Jones, and Lahey all attended the meeting at the DFW Airport Hyatt Regency, along with Vescovo, his CFO, Dick DeShazo, and attorney Matt Lipton. With Geffen tied up on another project, Atlantic's head of production Ian Syder, who would be handling the nuts and bolts of filming, flew over from London. Karen Horlick, who handled logistics for EYOS was also there, as was Phil Algar, whose company had been contracted to provide the ship's hotel staff. Vescovo had not invited Nekton's Oliver Steeds as he wasn't yet sure what incremental benefit Nekton could bring to the expedition and was already feeling saturated with relationships to manage.

McCallum had T-shirts printed with FIVE DEEPS EXPEDITION on them to help make everyone feel they were part of a team. He prepared a PowerPoint presentation listing the five objectives, describing how the schedule would work, and detailing how each group served the other.

The five objectives were straightforward: to execute all mission activities in a safe and efficient manner; to reach the five deepest points of the world's oceans; to undertake further exploration dives to targets of interest, including possibly the *Titanic*, USS *Scorpion*, and USS *Indianapolis*; to generate multidisciplinary science data for the global community, and to provide outreach material that will inspire and encourage further exploration. McCallum saw the film as part of the fifth objective, far down the list.

A lengthy discussion took place about installing a sonar on the ship to map the trenches. Everyone was advocating for the sonar, as it would increase the reach of the science, provide a better explanation of the ocean floor for the film, and add value to the ship itself. The price was a hefty $1.5 to $2 million for a state of the art multibeam echosounder made by Kongsberg, a Norwegian company, a number not at all contemplated in the ship's original budget.

"I was a big champion and so was Bruce of multibeam echosounder," Lahey says. "It was the only way you can substantiate the depth, and it's a hugely valuable asset to the vessel and an important part of attractiveness to an onward buyer."

Vescovo was annoyed at this very late, seven-figure addition to the budget seemingly out of left field. While Vescovo was wealthy, it irked him how so many members of the team—especially Jones—found it so easy to spend his hard-earned money and seemed to view him as an unlimited ATM. However, he eventually agreed that the logic of adding the sonar was sound—regardless of the messenger or tardiness of the proposal. The unanswered question became when and where could it possibly be installed without interrupting the expedition, as the ship would need to go into dry dock for several weeks so it could be attached to the bottom of the ship. McCallum, who had several shipyard relationships, began to look for a solution.

All in all, the meeting was a success—except for one exceedingly awkward moment. Jones announced that he and his wife would be diving in the *Limiting Factor* at the Challenger Deep. Lahey was aware of his partner's intentions, but had been so busy with the sub he hadn't addressed it. Everyone else in the room was surprised and looked directly at Vescovo. Inwardly, Vescovo was shocked. To him it was obvious that the two "Triton divers" should be Patrick Lahey—the builder of the sub—and, okay, maybe Bruce Jones, even though he was not directly leading a core part of the expedition any longer. More appropriate would be Lahey and the sub's gifted structural designer, John Ramsay. But Jones's wife? He thought she was a very nice person, but it just didn't seem fair or frankly deserving to Vescovo, compared to others who were contributing so much more blood, sweat, and tears. Jones's feeling was that Ramsay was young and would have other opportunities, and that it was Triton's dive so they had the right to decide the occupants.

With all eyes on him, Vescovo concealed his instinctive reaction and chose not to directly address the issue at that time. He needed things to keep moving forward, and with the exception of Jones, who had been moved to a nonoperational role, he was pleased with the team that had been assembled.

"It's a rainbow coalition on this expedition," Vescovo said after the meeting. "It's all about competence. It's not about any political agenda. When you are up on Everest and it's storming, you are saying to yourself, 'I'm glad I bought the best money could buy and I'm with the best team. When I'm down there at 16,000 PSI, I'll say I'm glad I spent the money I spent and have the best team."

CHAPTER 7
SWEATING OUT THE SUMMER

Victor Vescovo landed his Embraer Phenom jet at the small private airport in Vero Beach, Florida, on the morning of June 20. He had traveled to South Florida with his Caladan team, attorney Matt Lipton and CFO Dick DeShazo, so they could see for themselves the progress on both the *Pressure Drop* and the *Limiting Factor.* The ship, which was docked in Fort Pierce, was nearly ready to go, but the sub, which had been undergoing assembly for three months in Triton's facility, was behind schedule. It wasn't so much that Vescovo wanted to pressure Triton as it was that he needed to begin assessing the situation. The delays in assembling the *Limiting Factor* were threatening to create a major schedule overhaul to the Five Deeps Expedition before it even got started.

The pressing issue was that the sea trials in the Bahamas to test the sub needed to be finished in early August at the latest, or the *Pressure Drop* would not have enough time to steam to the Arctic Ocean for the dive at the Molloy Deep before the weather turned for the season. Missing the weather window would add at least three months and well over a million dollars to the expedition.

Vescovo, Lipton, and DeShazo rented a car and first drove to the port in Fort Pierce where the *Pressure Drop* was docked. Kyle Harris and his team were in the final stages of the refit. The ship had undergone a total transformation since any of the three had last seen her. When they pulled up, Vescovo commented, "Look how pretty she is, all new and painted."

Harris gave them a tour of the vessel, highlighting the work that had been done and outlining the remaining items. A large satellite communications dome that would provide Internet and telephone service for the ship was sitting on the dock. Along with the servers and computers, it was expected be installed the following day and brought online by Pippa Nicholas, the English-born communications expert.

"It's a wee bit shaky at the moment, but we will continue testing it," Nicholas told Vescovo.

Nicholas was one of what Vescovo called the project's "rainbow coalition." Since being hired, she had undergone transgender surgery and now liked to flaunt her new assets. The only thing that mattered to Vescovo, however, was that she was highly competent because comms issues at sea would be problematic for him personally. He needed very robust communications with shore to maintain his business interests while at sea.

Several other issues were being addressed, the biggest being the cradle and the hangar that would house the *Limiting Factor*. Originally, Triton had outsourced the design of the cradle that would hold the sub on the ship, but that hadn't worked. John Ramsay then redesigned the cradle and the hangar so that Triton could build and machine them in-house. But even when the setup was finished, unanticipated complications occurred when interfacing the hangar with the ship. Once again, something Triton thought would be easy was not.

The science freezers were running late because of miscommunications, but Harris was pushing to have them installed before the vessel sailed for sea trials. Harris had thought Alan Jamieson, the science team leader, had ordered them, while Jamieson assumed that Harris had. "The lines of communication haven't been the best," Harris said.

Overall, Vescovo was pleased with the ship. Lipton and DeShazo agreed that it had come a long, long way since they had last seen it. In the car driving to Triton, Vescovo told DeShazo that he wanted to give Harris a bonus for pulling the ship together in such a professional manner and on such a tight timeline. "I like to bonus people who deserve it and don't ask for it," he said.

When the three arrived at Triton, the mood was upbeat and anticipatory. Lahey shook hands with Vescovo and sarcastically asked if he would like to see the sub. "Do ya think?" Vescovo replied.

Lahey led Vescovo into the warehouse. He explained how what looked like pieces of a giant model would be assembled. Huge blocks of syntactic foam stamped with "389 kg" (858 pounds) were sitting on the floor. The thrusters (the sub's propellers and their casings) were spread out on workbenches waiting to be affixed. Vertical hoppers with steel slugs weighing 50 kilograms (110 pounds) each were off to the side, as was a bound mass of massive steel bars each weighing 250 kilograms (550 pounds) that would serve as the freeboard weight, the final weight dropped to make the sub fully buoyant. At the center of it all were the bolted-together titanium hemispheres that would house the pilot and passenger.

For Vescovo, just seeing all the components under one roof after three years of planning was exciting, even though nothing had been attached to the hull and the electrical components were scattered across several tables. With childlike enthusiasm, he climbed into the sphere to check on the height of the center console and get a feel for the capsule.

Next, he moved to a workbench with the manipulator arm clamped to it. The hydraulic arm, which would allow the sub pilot to interact with the atmosphere, had been manufactured by Kraft TeleRobotics at a cost of $350,000. They were the only company Lahey contacted that was willing to work on the project. It was the company's first product rated to full ocean depth and had the power to lift 180 kilograms (396 pounds). Vescovo grasped the handle. As if he were playing a video game, with the twist of his wrist he began maneuvering the arm and picking up wood blocks the way he planned to collect rocks and other materials from the bottom of the oceans.

Lahey explained that progress was slow because the company was waiting on numerous suppliers, many of whom were running behind, and consequently every day was one step forward, one step sideways. "We started to put the sub together in March, and we are waiting and waiting on all kinds of things that are behind schedule," Lahey said. "It's ridiculous the number of things that are behind schedule."

After the tour, the group went into the conference room for lunch. The current problem in a series of them was that the sub's batteries were delayed. Ictineu, the manufacturer located near Triton's Barcelona facility, was running behind, and the batteries would still need to be pressure tested once they were finished. Cost for 10 batteries: $1.5 million. "You'd think they could pick up that pace at that price," Lahey said.

Lahey had initially told Vescovo that Triton could obtain a permit in Spain to fly the batteries to the U.S., but that didn't pan out because the batteries are made of lithium polymer, a hazardous material. The options now were a commercial cargo ship or a private ship, or transporting them to another country and flying them in a private plane. Based on cost and practicality, the decision was made to put the batteries on a cargo ship to Miami and pick them up on July 19. "It's too bad my plane doesn't have the range, or I would go get them myself," Vescovo said.

Four days later, another distraction involving Bruce Jones reared its head. His son, Sterling, was let go from the ship because there was no real role for him going forward, and the following day he quit his job at Triton. Jones sent Vescovo an emotional email. "Empathy wouldn't appear to be something you do, and it certainly doesn't appear to ever enter into your decision making," Jones wrote. "You have managed to become vastly unpopular with both my wife and my youngest son; and you hardly know one another."

Lahey, already under enough stress, decided enough was enough. Calling his partner's email outrageous, he emailed Vescovo, "It would seem I will have to make a full time job of apologizing for my partner's behavior and all I can do is assure you this changes nothing as far as my commitment to you, the *LF* project and the FDE [Five Deeps Expedition]."

Given the fractured relationship between Jones and Vescovo, Lahey had come to the conclusion that it was best Jones not come on the ship at all. He told McCallum, who passed the news on to Vescovo. "I can't see him adding anything to the project other than stress and drama," McCallum emailed Vescovo.

For Jones, the end of his desire to be a part of the expedition would come a month later when Vescovo notified him that he would not allow his wife, Liz, to dive in the *Limiting Factor* on its discretionary dive at the Challenger Deep, even if Lahey piloted the sub. He said that it was an insurance liability.

Jones felt like the rug had been fully swept out from under him by Vescovo. "If you want to get to me, start with my son and end up with my wife and then you have an enemy for life," he says, punctuating his thoughts with an uncomfortable chuckle. "Victor is a great guy and I appreciate him as a client, but he agrees to lots of things then he changes his plan at the end. Guys who operate at that level in a dog-eat-dog private equity world try and create their own reality. That was pretty much the end of our relationship."

Vescovo, who no longer had time for the drama, says that Jones "did a series of things that showed that he not only needed to be out of the orbit, but in a different galaxy. He just hadn't delivered."

As June ticked away, the pressure on the Triton team increased as things were not coming together as planned. The fact that the company was making a first-of-its-kind submersible with no road map to rely on had become readily apparent.

In addition to suppliers running behind, every time a piece of the sub was affixed to the hull, something else had to be moved. The compensators needed to be repositioned because they did not fit in their planned locations. There was no space for all the electrical cables, resulting in the syntactic foam being cut. All the while, no one was quite sure how the batteries would fit into the junction boxes because they were still in production.

John Ramsay, the sub's designer, had relocated from the UK to Vero Beach for the final assembly phase. He was pulling his hair out over the problems. "I wasn't feeling overly optimistic," Ramsay admits. "Those last few months were utterly brutal. I didn't appreciate quite how much work had to go into the assembly and how different a new-build submersible was to what we had built before."

Admittedly, a major issue was poor project management. "We hadn't pushed vendors early on or got on top of production well enough to make sure everyone was going to deliver in our time frames," Ramsay says. "Every vendor pushed every delivery to the very last minute. Too much came in too late to make it a comfortable build. It was just unbelievably stressful."

Kelvin Magee, the veteran head of Triton's machine shop, was accustomed to the final weeks of assembling a submersible being chaotic, but he had never experienced anything quite like this. "Everything had to finish up at once," Magee recalls. "We were problem solving as we went. The confidence level that we could get the sub done in time for sea trials that summer wasn't very high. Patrick asked me one day, 'Are we going to get this done?' I said, 'I don't know, because I don't know what I don't know yet.'"

Lahey was dealing with vendors from all over the world: the U.S., Australia, Germany, Spain, the UK, and Canada. It seemed that every time he would give Vescovo a completion date, a vendor would notify Triton they were late, forcing the date to shift. When that occurred, Triton needed to re-plan what it could work on. The company was also forced to send employees on location to work with some of the vendors. One of the key engineers, Hector Salvador, spent most of his time at Ictineu dealing with the batteries at a time he was needed in Vero Beach for the build.

On June 28, Lahey informed Vescovo that the sea trials would need to be pushed back from July 16 to July 23. By July 4, it was becoming apparent that the July 23 date could not be met either. Lahey called McCallum to tell him the situation before he delivered additional bad news to Vescovo.

McCallum emailed Vescovo that "Patrick is sweating bullets and wanted to sound me out about another delay." He asked Vescovo to keep this between them so that Lahey would continue to use him as "a safety valve and confidant to bounce things off."

Buckle, the ship's captain, was overseeing the installation of the sub hangar on the *Pressure Drop* and the delivery of the support boats. Every time he interacted with Lahey, he could feel the anxiety. He told McCallum he was concerned about Lahey personally, which McCallum passed on to Vescovo.

Vescovo pushed back. "As for Patrick's mental state, well, I have to say he brought it on himself," he emailed McCallum. "Bruce is obviously of no help, and they have just taken on too much, and with too much optimism. But I can't afford to be overly sympathetic anymore. I have a lot of people and dollars at stake here. Patrick needs to go to war and get this done."

The following day, July 5, Lahey emailed Vescovo about the further delay. He admitted that Triton had fallen short of the deadline, but also made the case that Triton was not that far behind the original timetable. He pointed out that the construction contract for the *Limiting Factor* was signed on July 18, 2016, with an expected delivery date of 24 months later, which was to include sea trials. He assured Vescovo that Triton "has pulled out all the stops" working seven days a week, twelve hours a day.

"The *LF* is an order of magnitude more difficult and the other variables including the hangar, *LF* trolley, landers, and *PD* integration together have completely overwhelmed our capacity and used up our entire bandwidth," Lahey wrote. "Add to this the need for us to fulfill our obligations to other clients, which can at times conflict with our *LF* objectives, and you can appreciate the difficulty."

Vescovo was irked because regardless of what the contract said, Lahey had promised an earlier completion date upon which the entire machinery of the expedition—and its associated high expense—had begun. Lambasting Lahey over the multiple missed deadlines, he reasoned, wouldn't make things better, however. Instead, he opted for a measured, somewhat positive response. He explained to Lahey that "planning and commitments were made based on the schedule that Triton created, not Caladan or EYOS, that had us starting sea trials in early June, so in fact even if we make the July 25th, then by my reckoning, we will have had roughly a 7-week delay, not 17 days (as sea trials was supposed to be concluded in that 24-month period). But that will all be water under the bridge if we can make Molloy and *Titanic* in the critical weather windows with successful dives."

Copied on the email chain, McCallum wrote Vescovo: "Patrick is a dear friend but he needs to learn that optimism can be expensive. He needs to stop telling folks what they want to hear . . . and tell them as it is (with a contingency buffer!)."

Lahey admits it was a "clusterfuck," but adds: "Let's not forget we were building the world's first full ocean depth–certified human occupied vehicle."

Though Vescovo continued to believe that he had hired the best sub builder in the world and contracted with a first-class film production company, he was feeling frustrated at their ability to work together. There had been a miscommunication over the cameras, which were being supplied by the Woods Hole Oceanographic Institution (WHOI). Though Atlantic had been in touch with WHOI and the Triton team, somehow the final schematic for the cameras had not been given to Lahey until the week before departure for sea trials. "This is simply not acceptable," Lahey emailed Vescovo. Not only did there need to be an allowance in the sub's design for the cameras, the inspection agency required that they be independently pressure-tested to full ocean depth—not a short time frame activity.

By the time Lahey received Atlantic's drawing for placement of cameras inside the sub and outside the sub, it was too late to install them on the outside of the sub. A discussion ensued to see if temporary cameras could be mounted with Velcro for sea trials so that Atlantic could get underwater footage to show prospective broadcasters, but the Triton team had too many other items on its list to deal with the exterior cameras.

Vescovo emailed Geffen that the cameras could not be mounted in time for the sea trials dive. Geffen said that he understood and would discuss how to deploy cameras outside the sub on future dives. Vescovo was frustrated that he ended up in the middle of the situation that should have been taken care of by both sides.

∞

On July 12, Vescovo decided to make a surprise visit to Triton. After finishing a business meeting in the early afternoon in Richmond, Virginia, he and his copilot flew to Vero Beach instead of heading back to

Dallas as planned. Unannounced, he showed up at Triton to check on the progress of the *Limiting Factor.*

He walked into the lobby of the warehouse. Through the open door of the main high-bay, he could see the components of the *Limiting Factor* scattered around the floor, but there was no one working on the sub. The only work underway was on a different submersible.

Vescovo was not pleased. He had been promised by Lahey that the team was working twelve-hour days in shifts, breaking only for the heat. Lahey was at the facility, and the two talked briefly. Lahey attributed the inactivity to the late hour and the day's heat wearing people out. He told Vescovo that he was going to have a large A/C unit installed in the building to allow shifts to last longer. Vescovo was dumbfounded that something so necessary hadn't already been installed.

"The guys are busting their humps twelve hours a day, seven days a week," Lahey told Vescovo. "I've been burning my crew out every day. They are human beings, too. You have to let them rest."

The next day, Vescovo offered a bonus to Triton if it finished by July 25, as the financial consequences of not being ready were coming front and center. Lahey turned down the offer, almost as if Vescovo had insulted their dedication by thinking more money could make things go faster. But Vescovo had made his point: Triton needed to marshal all its resources to finish the sub sooner rather than later.

Still, on July 20 it was clear that Triton would not be ready in five days. Vescovo decided to give them four extra days and to pull back from his daily emails to Lahey. He felt that Lahey was so overburdened that any further prodding would only make things worse. He had made his point by showing up unannounced. Sometimes, he had learned, even in a battle you had to just step back and let things happen.

The batteries finally arrived from Spain that afternoon, as did the final cables necessary for other electrical connections. The last of the parts to finish assembly of the *LF* were due first thing the following morning.

"All major components now on-site," Lahey wrote to Vescovo. "It's going to be a busy weekend."

"Sounds like Christmas . . . and you even have the batteries," Vescovo wrote back.

A July budget analysis prepared by Dick DeShazo showed costs had escalated dramatically. The biggest jump was on the *Pressure Drop*. Originally estimated by Jones at $5 million on the high end, the final number was heading north of $12 million. This included the addition of the unplanned multibeam sonar to collect data to map the ocean floor, budgeted at $2 million. The *Limiting Factor* appeared to be coming in just above Triton's original cost estimates, but without substantial, customary overhead costs or any profit margin.

DeShazo estimated the total expected combined costs for the *Pressure Drop*, the *Limiting Factor*, and the around-the-world expedition to "shake down" and perfect the submersible would be over $40 million. Vescovo thought that given so much remained unknown and left to do, even that was conservative. He expected a full 20 percent more in unexpected costs would arise and that the entire system costs would end up being $50 million. To date, he had already spent more than half that amount. It was the cost of the remaining "unknown unknowns" that kept him up at night.

As the final budget numbers were being pulled together, Triton and Caladan were in the process of concluding the onward sale agreement for the *Pressure Drop* and the *Limiting Factor*. Each side was coming up with different numbers for the asking price, and there was a fluid debate over what costs would be added to the entire system to determine the sale price.

The key points had been agreed to. Vescovo only wanted to recoup his expenditures for building and perfecting the system and was not taking any profit. Triton would receive a reimbursement for foregone overheads and a minimal profit margin on the design and building of the *Limiting Factor*. If all went well, the company would independently receive from Vescovo a total of $2.5 million in bonuses, for DNV-GL certification, a successful dive to bottom of the Challenger Deep, and completion of the Five Deeps mission, which would be included in the asking price.

Another pressing question became what to do with the expedition costs, which were rising because of unaccounted-for line items and delays.

All estimates relied on the assumption that the expedition could make the weather window to dive the Molloy Deep in the Arctic Ocean in September—no sure thing because the *Limiting Factor* was still in pieces. If that window were missed, the Molloy Deep would have to be pushed to the fall of 2020. Because the expedition had been scheduled to end in May 2020 at the Challenger Deep, this would add another three to four months—a considerable additional expense.

Finding even modest sponsorship revenue to offset costs was going nowhere. Originally, Triton had estimated there could be $5 million in sponsorship income.

Nevertheless, Vescovo was comfortable proceeding, though at the same time he was trying to contain costs. In business ventures, he operated on the philosophy that if you can't control costs, everything else becomes irrelevant and you will never succeed. Uncontrolled costs equals venture failure—in anything, he had learned. Certainly, his abilities to control costs and maintain schedule discipline were being put to the ultimate test.

"Coming in 'over budget' by about 30 percent? Well, as these things go, I can't say it is a mortal blow," Vescovo wrote DeShazo in an email. "Could the U.S. government or Navy have done this for a similar amount? *No way in the world*. And why not? is the question that should be asked, and discussed. Just glad I am able to do it and hopefully we are successful."

❧

The next headache had nothing to do with the sub. Before sea trials, Vescovo was insisting that everyone sailing on the ship sign a waiver of acknowledgement of risk and limiting exposure to gross negligence in the event of injury. He was already fighting what he deemed a bogus workman's comp claim that had occurred during the refit in Seattle.

In light of that claim, the liability waiver was particularly one-sided. Participants had to agree to personally assume all risk for injury, illness, or death, and responsibility for any medical expenses that the participants or others incurred on board the ship. The waiver further stated the participant agreed not to sue Caladan "for injuries or damages whether they arise or result from any NEGLIGENCE or other liability, EVEN IN CASES OF GROSS NEGLIGENCE."

On the eve of sea trials, John Ramsay, the sub's designer, Tom Blades, the chief electrical engineer, and Richard Varcoe, the website designer, all refused to sign the liability waiver. When Vescovo found out, he pushed back hard, threatening to ask them to leave the ship and work remotely. DeShazo contacted Lahey, who tried to persuade Ramsay and Blades, both Triton employees, to sign the waiver, but they balked.

Faced with a decision, Vescovo decided to allow the three on the ship for sea trials and discuss the issue with them. There was simply no practical way to conduct sea trials without the sub's designer and the electrical engineer on board. He related to them his experience with a workman's comp claim already in motion, and he wanted them to commit to him that they wouldn't file another. Enough to his satisfaction, they did, and they were allowed on board.

⁂

With the clock ticking and the Molly Deep weather window closing, Rob McCallum flew to Florida to monitor the situation firsthand. On July 25, McCallum arrived at Triton around 6:30 A.M. with bagels and coffee to check on the final assembly of the *Limiting Factor*. He felt that the tide was finally turning, and that completion was in sight.

"(The extra four days you gave them) has served to lift them out of the 'fug' and into a more positive and energized run for the finish," he reported to Vescovo via email. "There was activity in every part of the building today; with John (Ramsay) and Jonathan (Struwe, the certifier from DNV-GL) going through test protocol, Richard on the website, 4 different benches of electrical work, 3 folks working on the dummy and 6 on the fitting of lights and external hardware. The 4 days delay has been a good investment and in hindsight they were not going to make tomorrow."

Two days later, on July 27, McCallum met with the Triton department heads individually. After hearing from each one of them, he concluded that mobilizing for departure to the Bahamas the following evening was not possible and asked each one for an honest appraisal. A consensus was reached that Thursday, August 2, could be met.

An understandably impatient Vescovo was miffed. The topsy-turvy scheduling was wreaking havoc on his scheduled workload at Insight Equity. He and his partners were in the process of selling a construction company that he was chairman of and which they had owned for ten years, but there were numerous technical wrinkles. He was also in the midst of a potential $80 million purchase of a lumber company in East Texas. Complicating matters, however, was the owner. A good old boy, the man had declared that he only did business on a handshake, and thus refused to let the Insight team talk to any of his customers. With every delay, various meetings with important people had to be rescheduled.

Surprisingly to some, Vescovo hadn't even told his partner, Ted Beneski, whom he had worked with since 1994, about the *Limiting Factor* or the expedition. "He doesn't tell me what he does on his weekends," Vescovo said. He planned to break the news to Beneski the day before it became public. Another, less optimistic and darker part of him wondered if the sub would actually be built, test successfully, and the expedition ever started. He decided he would announce what he was doing to others when it was really "on."

He changed his departure plans for the Bahamas a third time. "The general verdict? Patrick is a great sub pilot and builder, but he is not the best project manager," he said. "He has a poor habit of letting his optimism get in the way of cold calculations, and leaving little or no allowances for the natural friction of business, tech development, or construction. That isn't a damnation of him, it's just that we've discovered his weakness and it is a big one that I wish I had known sooner. I could have helped mitigate it with more milestones and inspections. I should have known better when they didn't deliver the first simulator on time—it was more than a month late. People that miss schedules don't usually get a lot better at it with larger and more complex things."

August 2 came and went, and on August 3, McCallum reported to Vescovo that there was a further 24-hour delay in loading the *Limiting Factor* onto the *Pressure Drop*. The repeated delays had left no time for harbor trials in Fort Pierce, meaning that the sub would not enter the water until it arrived in the Bahamas. That violated their original plan and skipped a whole "harbor trial" testing interval.

That evening, McCallum left Ft. Pierce to run a previously planned expedition in Svalbard, Norway, somewhat ironically the place the *Pressure Drop* needed to be in less than a month. The prolonged building period was causing the expedition leader to miss sea trials. He put his capable colleague Richard Bridge in charge, but not having McCallum there was less than ideal given the intense state of play and his detailed involvement.

On August 3, the *Limiting Factor* passed the Factory Assembly Testing, an inspection conducted by DNV-GL's Struwe, who would accompany the Triton personnel to sea trials. The team had been in constant contact with Struwe throughout the design and build phases so that there would be minimal problems with certification. The next step would be for Struwe to observe the sub while it was operational and rate its capability and most importantly, its safety.

Finally, on August 4 at 11:30 A.M., the *Limiting Factor* was trucked from Triton's facility to the port in Ft. Pierce. The submersible was loaded onto the *Pressure Drop* that afternoon. As the ship sailed for the sea trials in the Bahamas, the Triton team was still doing the final assembly on the sub. "What I didn't want to happen was to finish the sub when it was on board the ship, which is exactly what happened," Triton's Magee said.

Magee had several concerns, notably the durability of one of his final-hour fixes. There had been problems sealing the trunking that the pilots would use to enter the sub and enter the titanium sphere. Magee had come up with a decidedly low-tech solution—he had used a Schwinn bicycle inner tube as the rubber seal.

"My biggest fear was that the trunk was going to leak from the bicycle tube seal," Magee said. "I also never trusted the frangibolt holding the batteries or the thrusters. Each battery packs weighs 400kg. The manipulator arm weighs 100kg."

Because of the manufacturing delays on the batteries, Triton had only the six batteries needed to run the sub and no spares. If a battery were lost or malfunctioned, the sea trials would be over.

Relieved and feeling optimistic, Lahey sent a heartfelt email to everyone involved in assembling the *Limiting Factor*, calling it the most important project in his life. In part, he wrote: "I wanted to take a moment to remind all of you that the journey you are a part of is historic. The *LF* project is bigger than all of us. This submersible promises to really make a difference and allow human beings to regularly and safely visit the deepest, most remote and least understood part of our planet for the first time in history. We should all step back, catch our breath and consider the magnitude and significance of what we are doing right now and revel just a little bit in the moment. Life is fleeting and not everyone has the opportunity to be a part of a project that can move the needle, shape or impact history and change people's perceptions forever. You all have this opportunity and I do hope you are as proud and excited as I am to play a role in this remarkable endeavor."

As inspirational as Lahey's email was, there remained a very big unknown: because the sub had never been in the water, no one was exactly sure how it would perform. Vescovo's feared "unknown unknowns" would soon surface.

CHAPTER 8
SEA TRIALS

The *Limiting Factor* wasn't performing. The shiny, white two-person submersible, measuring 15 feet long, 6.2 feet wide, and 12 feet high, weighing 25,700 pounds, or 12.8 tons, might have been a feat of inspired design, technical engineering, and electronic wizardry, but from the moment the first-of-its-kind submersible was lowered into the water, it was also clear that the *Limiting Factor* had its share of gremlins.

The sea trials were being held in the Bahamas, a couple miles off of Abaco Island. The location was chosen for the azure blue, calm waters on its westward side, a place to give the Triton Submarines team optimal conditions to test the launch and recovery of the *Limiting Factor*. Less optimal was that the owner would be one of the sub's test pilots and a film crew from Atlantic Productions would be on board the *Pressure Drop* filming them.

Rob McCallum, the expedition leader, didn't want Vescovo or the cameras near the sea trials, but that wasn't in the cards. "I made very clear that it was a high-risk time," said McCallum, who missed the sea trials because of a prior commitment. "Rule number one is you need to keep the client away from the sub until it is working as planned. But because Victor was the owner and one of the test pilots, that was the beginning of the frustration that he started to experience—'I've got this new toy, I paid a lot of money for it, I'm hugely invested in it, it's my baby, and it's not working like I think it should.'"

Normally, Triton would have ironed out the kinks before letting the client dive in the sub. And even if a film were being made using one of their subs, they wouldn't let the cameras near it until the sub was fully operational. But in this case, the client was essentially managing the entire project, which was running so far behind schedule that the expedition might have to be rerouted and extended, and the film crew needed footage to secure a broadcaster and its all-important funding.

Atlantic had six people on the sea trials. To capture the footage, it used one camera on deck, one in the tender, two inside the sub, and a drone. Additionally, a diver would use an underwater camera to film the sub descending.

Triton's Magee, who has an easygoing demeanor, was less than pleased with having a film crew on board. "We are working with a brand-new sub and a brand-new boat while TV people film us going through our learning curve," he said. "We're doing it with our skirt right up to our ears. I think having the film crew here was the wrong thing to do. We can't admit mistakes and have our usual testing process because it will end up in the film. It's not how you do things. We don't go out with our client to test his sub, and we sure as fuck don't go out with Atlantic Productions."

Vescovo certainly understood this, but was between a rock and a film camera. Without any video of the system in actual operation, Atlantic was likely to abandon the project. He would then have to fund the multimillion-dollar budget to properly film the expedition for posterity, or scrap the filming of it altogether. Neither was an appealing option, so he decided to endure the displeasure of Triton instead. It was the least worst option at that time.

Initially, Triton had planned to conduct harbor trials in Fort Pierce. This would have allowed it practice the Launch and Recovery System (LARS) and to test other deck equipment and systems aboard the *Pressure Drop* that were directly involved in submersible operations, including the lifting winch that picked up the sub. These systems included the tag line winches that stabilized it, the towing winch, the "knuckle boom crane" that launched the *Learned Response* support boat, and the A-frame crane that lowered the sub into the water. But they had run out of time, so all of these would be tested for the first time in the Bahamas under the glare of the cameras and the scrutiny of the owner.

Lahey was tense, but trying to be circumspect about the setup. "Sometimes in the rush to completion you don't always do things the way you might have liked to do them, you do them in a way you needed to get the job completed," he said. "You make compromises as you have to, though not ones that jeopardize the vehicle's safety. Everyone knew going in that it was unrealistic to have everything working perfectly, and we would just work through them in a more public forum than we would have normally liked."

To no one's surprise, problems occurred on the first dive. Lahey piloted the sub with Magee riding shotgun. The waters were calm, the weather ideal. They dived the *Limiting Factor* to twenty meters to test its core systems. The first thing they noticed was that the main ballast tanks that were supposed to fill with water to cause the sub to submerge weren't sealing properly. This meant the sub would not descend properly.

"There are 'duck' valves that are supposed to allow water to flow in but not out," Magee explained. "They failed miserably, so we have to redesign them."

A bigger issue was the *Limiting Factor*'s buoyancy level above the water's surface, known as its "freeboard."

The sub surfaces by dropping weights. The first weights dropped are the sixteen 5-kilo (11-pound) Variable Ballast Tube (VBT) biodegradable steel weights, which are dropped one by one about 200 meters from the bottom until the sub is neutrally buoyant. The VBT weights are housed in two long columns on either side of the pilot sphere and dropped sequentially. When the sub is ready to surface, the pilot drops any remaining VBT weights and also a single, 250-kilo (550-pound) surfacing weight which causes the sub to ascend fairly rapidly, about 2.5 feet per second. Once it breaks the surface, a large freeboard weight is dropped, pushing the sub high enough for the majority of the trunking column to be above the waterline. Most of the water in the trunking is then pumped out mechanically. After the sub is attached to the ship by a towline, before the hatch can be opened, a "swimmer" enters the trunking and pumps out the remaining water with a handheld pump. The swimmer then opens the now dry hatch so that the pilots can exit the sub.

But when the sub surfaced on its first dive, and then again on its second dive, it wasn't becoming positively buoyant, meaning its freeboard was too low. This meant that water was sloshing into the trunking faster than the pumps, or even the swimmer, could pump it out, and therefore opening the hatch was difficult. More concerning was that this was occurring in calm waters.

"The fine details of its buoyance couldn't be worked out until the sub was in the water because we didn't know exactly how much it weighed," explained sub designer John Ramsay, who was on board despite his refusal to sign a liability waiver required by Vescovo.

Jonathan Struwe, the DNV-GL certifier, was on the ship observing and offering guidance. Because of the buoyancy problems, he could only rate the sub to sea state 1, pejoratively called "mill pond" conditions. It needed to be rated to at least sea state 3, moderately rough conditions that would be faced on open water in the oceans.

The list of gremlins stacked up quickly. The manipulator arm was not operational. Semiconductor fuses failed during the on deck start-up procedure. The starboard cameras were not working. The battery pods need to be changed too frequently. And the launch and recovery of the sub was a shaky procedure at best.

"We need to figure out how to get the thing over the stern and back on board before anything else matters," Magee said.

A side issue was that the ship's crew struggled to deploy the *Learned Response*, the protector boat that triangulated communications with the sub and the ship. The boat was launched from the second deck using the knuckle-boom crane. The process went so poorly the first two times that Captain Stuart Buckle told Lahey that the boat would have to be replaced, as it appeared that it could not be raised or lowered on its cradle in anything greater than sea state 1.

Vescovo, whose arrival at sea trials had been delayed by Lahey in hopes that the sub would be fully functioning before he arrived, took the problem in stride.

"They bought a protector boat that is just too big and ungainly to be effectively used where we are going," Vescovo said. "So now we have to figure out what kind of boat we can use effectively and maybe even have to do structural modifications to the *Pressure Drop* to accommodate a different boat or to ease launch and recovery. This problem will get solved too, but will also take time and also money—not to mention potentially $300,000 wasted on a boat I can't use."

Buckle was also concerned about the lack of safety procedures being followed. The Triton crew didn't all wear helmets, and some wore flip-flops instead of protective shoes while working on deck. At one point, Lahey was nearly decapitated by a flying metal hook. "My concern is the safety of everyone on board, and they are being very cavalier about it," Buckle said.

And only one of the three landers was operational. With so many problems, the chances of making the Molloy Deep weather window in 2019 were rapidly starting to fade.

"It isn't the end of the world," Vescovo added. "The basic design is still valid and I am confident the sub can be 'fixed,' but killing Molloy in 2018 will add another two to three months to the expedition and cost at least $1 million in crew expenses."

The good news? "Everyone says the food is great," Vescovo quipped.

⁂

The sea trials were the first time the film crew and the sub team had met one another, and the Triton team felt like it was under a magnifying glass. The Atlantic team was concerned at how many things weren't working with the sub, as other expeditions they had filmed were always up and running by the time they arrived. But all other expeditions had the luxury of precedents and sea trials conducted before the unveiling of the sub. The *Limiting Factor* was not only 100 percent new and in virgin territory, it had not even been in the water before sea trials because of time constraints.

Though both Triton and Atlantic were under different pressures, their goals were mostly aligned. Triton needed to test and troubleshoot the sub and ensure that the LARS (Launch and Recovery System) worked. Atlantic needed to film enough footage of the sub in action to convince the Discovery Channel that the sub was viable. Network executives would understandably not sign off on funding a series until they saw evidence that the sub could dive the ocean depths, which meant that Atlantic had to fund the sea trials shoot.

In an attempt to bring a production deal together quickly, Anthony Geffen, Atlantic's CEO and founder, had invited a Discovery Channel executive to watch the sea trials. Geffen, who had already been forced to eat the costs of having a crew ready twice before when the arrival of the sub was delayed, was understandably eager to secure Discovery's commitment.

Atlantic had made films of underwater exploration all over the world, including two on billionaire Ray Dalio's ship using Triton subs, but Geffen was concerned by watching the operations on the *Pressure Drop*. It took three days for the Triton team to get the sub in the water, and the results on the first dives were underwhelming to say the least.

Indeed, the Discovery executive, who watched the first two days, called the situation "chaotic." No one involved could disagree.

∞

After receiving the reports of the problems with the sub, Vescovo finally decided he needed to see what was happening firsthand. Rather than wait until Lahey informed him that the sub was fully operational, he made plans to fly to Marsh Harbor in hopes of enough issues being resolved that he could dive the sub.

Vescovo's plan was to fly from Dallas to Jacksonville with his copilot, Manny Montes, on August 8, overnight there, and then head down to Marsh Harbor first thing the morning of the 9th.

He took off from the private airport in Addison, Texas, in the late afternoon of the 8th. Somewhere over Alabama, he and Montes were reviewing the checklist of documents needed to enter the Bahamas. One key document was missing: Victor's passport. "No, really, I'm not kidding," he told his copilot. In a hurry, he had left it in his car when switching luggage, which was parked in his airplane hangar.

With no other alternative, Vescovo radioed air traffic control that he was returning to Addison "for operational requirements," and literally turned the plane 180 degrees around at 41,000 feet. An hour later he was back on the ground in Addison. After refueling, passport in hand, he took off again for Jacksonville. The false start somehow seemed fitting.

The following morning, Vescovo and Montes flew to the tiny airport on Abaco Island. He touched down in the late morning and was driven an hour to a dock where he boarded the *Learned Response* to take him to the ship. Geffen and his camera crew were on the support boat filming Vescovo's arrival.

"I thought it would be fun to maybe be the first guy to fly a plane and dive a sub on the same day," Vescovo commented offhandedly. "How many people have done that?"

Vescovo boarded the *Pressure Drop*. Lahey greeted him and told him that the plan was to do two more practice dives that day before Vescovo dived the following day.

"I want the sub to be working and ready before you dive it," Lahey told him.

Vescovo was clearly disappointed by not being able to dive on the day he flew, but he deferred to Lahey.

The two reviewed the list of issues, which also included the battery banks not functioning properly.

"Given all these issues, do you think we will be able to do Molloy?" Vescovo asked.

"I do, barring some unexpected problems," Lahey said, ever the optimist. "The next couple days are critical."

∞

With Vescovo observing, Lahey and Magee dived the sub. The dive was more successful than the previous ones, but there was a close call when the sub resurfaced. It came up about fifty feet from the ship, far too close for comfort. There had been a disconnect in communications between the *Pressure Drop* and the *Learned Response*. The *Pressure Drop* control room had said not to bring the sub up, but the *Learned Response* crew cleared the sub to surface.

"We popped up way too fucking close," Lahey said when he was back on board. "We need to work that out."

Several electrical problems also persisted. Lahey vowed to work through the night to correct them so that Vescovo could dive the following day.

After the Triton crew had worked through most of the night and much of the following day, the sub was ready for Vescovo to dive it. He and Lahey boarded the sub for his first dive.

The launch and recovery went smoother than the previous dives. As the sub descended, Vescovo turned to Lahey and told him, "I love the sub. Thank you."

The dive was both to test the sub and for Vescovo to become comfortable at the controls. He had spent more than forty hours working with Lahey via Skype in a simulator in his garage in Dallas. Being a fixed wing and a helicopter pilot, much of the process, like constantly checking oxygen and CO_2 levels and monitoring electrical systems, came naturally.

Vescovo and Lahey stayed down ninety minutes. Though the sub performed reasonably well, there was a laundry list of things that did not function properly. Vescovo's headset did not work, nor did the rear camera, which meant that many of the circuit boards would need to be pulled and tested. Intermittently, the sub's headings displayed on the screen did not change when the sub changed directions. One of the batteries, which has 160 volts, showed 0 on the monitor, despite the fact it was providing power. The readout on the VBT weights—those released at the bottom to make the sub neutrally buoyant so it can hover—showed they had discharged, even though they hadn't. Several alarms were also going off that shouldn't have been.

Still, back on deck, Vescovo was visibly excited by the experience. It had been 1,429 days since he had first emailed Triton about building him a submersible, and he finally dived it. "The dive was awesome," Vescovo said. "We have several things to work out, but the core systems are working."

Lahey was relieved that Vescovo finally got to dive the sub. "It has been a rollercoaster ride," Lahey said. "The team is exhausted from last three months but has also been invigorated when you actually see it happen. The sub is not 100 percent, but maybe 90 to 95 percent. We've had some big setbacks, some where people would be inclined to throw up their hands. And then we had some big victories. Now we are going hell for leather until we are over the finish line."

The biggest problem to date came to light in the ensuing hours when one of the landers failed to resurface. Two landers, *Skaff* and *Flere*, had been deployed before the dive. *Skaff* resurfaced on cue but *Flere* did not.

The landers were designed with two layers of redundancy so that they would resurface. The first means of release was actively "pinging" the lander acoustically through its modem to drop its surfacing weights. If that failed, the backup system was an "egg timer" that was set to release the weights at a certain time after launch so that the lander would resurface. However, neither release had worked. The lander was still on the bottom of the seafloor. Where exactly, no one knew. Vescovo was surprised that there had only been a double-redundant, and not triple-redundant, system on something so valuable.

A discussion ensued about what could be done to find the lander, as a new one would cost in the neighborhood of $300,000—more than the entire sea trials. Tom Blades had the coordinates where the lander had been deployed, but there was no way to know if it had drifted. Lahey suggested they dive the *Limiting Factor* and see if they could spot it. Hopefully, the lander's lights would still be working.

Lahey also got in touch with Triton client Carl Allen, who owns a 3,300/3 sub and the support vessel *Axis* and asked about using his sub to search for the lander. Allen okayed the mission. Lahey made plans to send two Triton employees, Mat Jordan and Frank Lombardo, to meet the *Axis* in the coming days and work with Les Annan, the *Axis* captain and an accomplished sub pilot, to find the lander.

That night, Lahey and Magee dived the sub again so that Atlantic could film it glowing under the water. Geffen sent a diver in the water to film the sub from below. He had also been allowed to attach temporary

Go Pro cameras to the sub. After the night dive was complete, Geffen now had enough footage to cut a teaser reel.

The following day, before Vescovo departed the ship, he and Lahey dived the *Limiting Factor* again, this time to 480 meters, the deepest it had gone. The launch was dicey, and the recovery took more than a half hour, far longer than expected. The issue was that even in sea state 1, the sub rocked back and forth. While sacrificing surface comfort had been a design choice Ramsay made to make the sub faster when going up and down in the water column, in a more harsh sea state the pilots would be uncomfortably tossed around, particularly if the recovery process was not more fluid.

Ramsay was frustrated that the sub wasn't ready for prime time. "It's a bit embarrassing," he said.

By the final day of sea trials, Vescovo coldly concluded there were just too many problems to proceed to the Molloy Deep and believe that they would be successful. The plan now was to send Triton back to Vero Beach to execute the punch list and then return to rougher and deeper water in the Bahamas for what was being called "*Advanced* Sea Trials." This would cause the expedition to miss the Molloy Deep weather window, thus pushing the Arctic Ocean deep off until the late summer of 2019. And so, before the Five Deeps Expedition had even started, it had already been extended by at least three months.

"You've got to hand it to Victor, he can take the knocks, he really can," McCallum said. "He takes the bad news well."

Geffen was pushing Vescovo to kick things off at the *Titanic*, rather than start with the first—and much closer—Puerto Rico Trench off Puerto Rico. He had tentatively sold the idea for a one-hour special examining the deterioration of the wreck to National Geographic Channel and other broadcasters, as Discovery Channel had turned down doing an additional show beyond its possible commitment to the Five Deeps series. The *Titanic* dive would be the first manned dive to the wreck since 2004 and the first that would return 4K images.

"Victor, *Titanic* is huge," Geffen told Vescovo over dinner on the ship. "It's the perfect launch for the Five Deeps."

Vescovo agreed, but weather again was the issue. Depending on how long Triton needed before Advanced Sea Trials, a *Titanic* dive would not likely be until late September, or even early October, at the tail end of the ideal time to dive there. Over the next few weeks, Buckle and McCallum weighed in. Both agreed that it *could* be done, but of course, there was no way to predict the weather off the coast of Newfoundland.

"*Titanic* is riskier in October, but we can wait in St. John's (Newfoundland) for a weather window and scamper out there," McCallum emailed Vescovo. "There is a higher chance of bad weather (Autumn is coming), but before it comes there are usually some nice calm days and settled periods. It is not ideal, but the fact is that I simply do not have any faith in Triton's dates . . . so I want to give them more space than they would ever ask for, and then work with you to drive them hard to perform and deliver."

~

Aside from the *Limiting Factor* not behaving during the first sea trials, the teams were operating as factions focused on their own interests, not as a single unit as McCallum had preached at their pre-expedition meeting in Dallas. The strife spilled out into the open after the sea trials ended.

Lahey was single-minded about Triton's interests. In addition to making the *Limiting Factor* operational, he was advocating that the *Learned Response* did not need to be replaced. If it were, the cost of the new boat would be added to the expedition cost, thereby decreasing Triton's potential level of cost reimbursement. He blamed Buckle for not paying close enough attention to the launch and his crew for not rigging the boat properly.

"I watched the crew aboard *PD* attempt to launch and later recover the *Response* and it was clear they did not know what they were doing and had not thought through the process at all," Lahey emailed Vescovo. "I saw men holding lines while being catapulted through the air and slammed into hard structures on deck."

Lahey was also irked by Buckle throwing out several negative jabs about the *Pressure Drop* in front of his crew. Lahey, who along with Bruce Jones had selected the ship, was already sensitive about the ship because of the soaring costs of the refit and had taken to trumpeting its better characteristics, namely that it was roomy, had minimal acoustic noise, and was extremely fuel efficient.

"I remain concerned about the constant flow of negative comments from Captain Stu regarding *PD*," Lahey continued in his email. "He continues to characterize *PD* as an old, tired and worn out platform, which is tedious and destructive. *PD* may not be a state of the art [ship] like the other vessels Stu is accustomed to operating but she is a fantastic and capable vessel ideally suited to her current role. If Stu has such a low opinion of *PD* and her capabilities, why did he take the job of Captain?"

Vescovo asked Buckle about his making derogatory comments, which could be damaging to morale as well as the onward sale of the vessel, but Buckle denied making any negative comments and said that he would not do so in the future. As the gripes poured into Vescovo's email inbox, he turned to McCallum to be peacemaker. "Hopefully he can help short-circuit what appears to be growing hostility between Stu and Patrick," he said of his expedition leader. "It isn't helpful, like at all. Both sides keep making snide comments about the competence of the other. I personally think there is some truth to both sides, but much less than is real."

McCallum stepped into the breach as the arbiter. "Patrick is likely feeling that he's lost ground and credibility," McCallum told Vescovo, "I'll work on illustrating to him that the key to success is the success of the team. There cannot be factionalism. Not now, and not when it gets tough."

Sea trials had become people trials. In the days following the first sea trials, another tense back-and-forth email exchange took place concerning the lost lander. Theories abounded as to what had happened to *Flere* and who was to blame.

The L3 Communications team that operated the lander's modems faulted the lack of a launch checklist and said that the crew had failed to attach the power cables from the batteries to the electronics, evidenced by the fact that they had never had contact with the lander after it was deployed. While the battery-driven "egg timer" should have released without the power connection, they also said that two of the four egg timers they had used had previously failed, so they were going in with a 50 percent failure rate on the backup system. Triton was planning to return those egg timers and find new ones after the sea trials, but it appeared that the damage was done.

Lahey told Vescovo that he believed the lander had flipped over. "The terrain at the site is rocky and rather steep for a lander and it is entirely possible *Flere* struck a boulder or haystack on the way to the bottom and toppled over and headfirst into the bottom rendering the modem and release mechanisms unable to work properly," he said.

Alan Jamieson, the chief scientist, rejected Lahey's theory outright. He explained that as a matter of physics the weight and buoyancy distribution made the possibility of the square-shaped lander flipping over pretty much impossible. Despite the fact that Jamieson was supposed be in charge of the landers, he had let Triton design them. He blamed Triton's construction of the landers, which he pointed out was already suspect because the electrical board that Triton had installed in lander *Closp* was built wrong and it had fried the moment the power was turned on.

"It became apparent that nobody fired up the modem on deck prior to launch, therefore I think a dead modem is more likely the result of connector damage through water ingress the day before," Jamieson emailed Vescovo. "I can't reiterate enough that the landers cannot end up upside down or on their side, they are only 40kgs in water with loads of buoyancy pulling up and the ballast pulling down. It just can't happen."

Vescovo sat back and watched it all play out. "Patrick was afraid I was going to deduct dollar-for-dollar from my final payment for the *LF* the cost of the replacement lander, as well as the cost to extend the FDE by another two months ($500k?)," he later emailed. "Dr. Jamieson was thinking I was going to fire him. And the calming influence of Rob McCallum was out of the picture. Anyway, welcome to the circus. <wry half smile>."

A week later, *Axis* sailed to the Bahamas with Triton's Lombardo and Jordan on board to search for the lander *Flere*. They rigged a system to pull the lander out the silt if they located it, but after a day of searching for it in an acrylic Triton sub, they had no luck.

Lahey delivered the news to Vescovo and said that he had implemented a plan to build a replacement lander, dubbed *Flere 2*, by the second sea trials. However, he cautioned that it would be a challenge to get the syntactic foam from the UK, a new modem, and a replacement junction box with all the required cables and connectors in such a short period of time. He also asked Vescovo to pay the expenses of the *Axis* team for the search.

Vescovo pushed back slightly before agreeing. He asked Lahey if he had offered a "bounty" for the lander in the event that it had surfaced after the ship left the area. While Abaco Island was relatively remote, the other areas of the Bahamas were heavily populated with tourists and boaters. He also floated the idea of sending his helicopter to the area to conduct a detailed, low-level search along the shoreline.

Vescovo told DeShazo and Lipton that Lahey "offered his apologies for losing the lander, but it rang a bit hollow to me, truth be told. It kind of felt like, 'Oh well, sorry about that. Please send another check for another lander.'"

He was more concerned that Lahey had gone forward with sea trials with a sub that wasn't ready. "He went to sea with major systems not operational," Vescovo said. "He is the kind of guy that looks up at Everest from base camp and thinks to himself, 'Oh, that should only take a couple of hours . . .' Inexperience and unabashed optimism is a bad mix here, but we are slowly getting a handle on it."

After the sea trials, Vescovo prepared a detailed "anomaly" list of more than fifty items that needed to be addressed before Advanced Sea Trials—"again, dragging the donkey out of the mud," he said. He put them in three categories: one for "must-have" items, two for "could-wait" items, and three for "when you can" items. Among the eleven category one items that needed to be addressed on the *Limiting Factor* were the duck valves, the VBT system, which sometimes jammed and did not release the weights properly, and multiple alarms that went off because of improper power distribution. Category two included labeling the light switches and fixing the A/C system. On the Toughbooks, category one items included fixing the descent rate, the heading indicator and the depth accuracy on the surface, as the GUI showed the sub 10 to 16 feet underwater when it was on the surface. There was a separate list for the landers. Top priority: a triple redundancy recovery system to prevent the loss of another $300,000 lander.

Lahey originally targeted August 27 for the Advanced Sea Trials, but Vescovo had little confidence that Triton would be ready by then. He told Lahey that he would not green-light a second sea trial until all the category one items had been checked off the list. The target completion date was moved to the first week of September.

On Labor Day, September 3, two days before the *Pressure Drop* was scheduled to sail to the Bahamas, Vescovo made yet another trip to Triton to check on the progress of the anomaly list. "I feel like Darth Vader going off to inspect the behind-schedule *Death Star*," he joked.

The first thing he noticed when he arrived, which slightly annoyed him, was that Lahey had given everyone the day off. Sure, it was Labor Day, but the *Pressure Drop* was just forty-eight hours from sailing back to the Bahamas for a second sea trial, which needed to be successful before an attempt was made to dive the *Titanic*. Lahey brushed it off, saying everything would be ready by launch time.

Vescovo spent nearly three hours at Triton's facility meeting with Lahey, as well as with electrical engineer Tom Blades and chief scientist Alan Jamieson, who was there to supervise the rebuilding of the landers.

Big-ticket items such as power distribution, the VBT system, and the duck valves were finished, and the Triton team was deep into the other category one items. Vescovo and Lahey did a full start-up test of the *Limiting Factor* in the shop, and all systems worked as they should. They had a "clear electronics board" for the first time. Two of three cameras were working, with the aft camera set to be installed the following day. But several analog instruments required by DNV-GL for temperature, humidity, and barometer still needed to be hooked up, and the manipulator arm was still not active.

Jordan, the Triton electrical engineer, had designed a more robust timer circuit for the landers. In addition, an 800-pound breaking strength, galvanically corroded release set at twenty-four hours would be installed as the third redundancy that Vescovo now required.

On the electronics front, Triton was also waiting for DNV-GL to approve the underwater cameras that Atlantic Productions was licensing from the Woods Hole Oceanographic Institution before installing them. There were so many electronic issues that "Blades was in 'an execution bottleneck' even though Patrick has a hard time admitting it," Vescovo said after his visit.

Lahey had been through the building and testing process many times and was sanguine about it. "I've never encountered any fucking sub project that I have done in my fucking nearly thirty-eight years of doing this stuff where it is easy right out of the gate," he said. "At the beginning, it's brutal and it's going to look like a train crash in slow motion, and that's exactly what it does look like. But eventually, you figure it out."

⁂

The *Pressure Drop* sailed for round two of sea trials in the Bahamas on September 5. A series of dives over the next three days showed remarkable improvements. On its first dive, the sub maneuvered and performed to expectations, with only a few electrical glitches. The following day, it dived to 1,180 meters. Then, on September 9, Lahey and Struwe went to 4,926 meters, a depth that ensured DNV-GL would commercially certify the sub, and a dive that showed depth wasn't going to be an issue for this vehicle. It was functioning like a tank underwater.

On each dive, the LARS team was becoming gradually more efficient at launching the sub and bringing it back on board, but there was still work to be done before heading into rough waters. "Our performance has gone from a C to a C+," McCallum told Vescovo.

Vescovo arrived on board the evening of the 9th. The relatively successful deep dive had moved the needle on the Triton's mood from frustrated to cautiously optimistic.

The following day, Vescovo dived with Lahey. They targeted 5,000 meters. The launch went off without a hitch. As the sub descended, despite a few minor issues, the two had strong communications with the ship and the dive was going well.

But when they reached the bottom, at 4,900 meters, and Vescovo turned on the hydraulic system and attempted to use the manipulator arm, it abruptly shut down. A small puff of smoke wafted through the cabin. A fire in a sub is one the worst things that can happen because it eats up oxygen rapidly, and to put out a fire the pilots either have to contaminate the air with fire retardant or take out the oxygen—both of which are harmful to humans.

Vescovo and Lahey looked at each other wide-eyed and were silent for a single, long second.

"Do you smell that?" Vescovo asked.

"Yeah," Lahey replied.

"What was that?" Vescovo asked.

"I don't know," Lahey replied.

They pulled out and readied the scuba regulators for emergency breathing and then began double-checking all systems to determine if there was any imminent danger that needed to be dealt with. There was no depth reading, as the CTD circuit responsible for providing the reading was apparently fried. The manipulator current connector also appeared to be fried, but nothing appeared life-threatening. A trained pilot, Vescovo did what any pilot would do, he undid the last thing he did before the smoke appeared: he turned off the hydraulics. The smoke did not reappear.

They signaled the *Pressure Drop* that they were immediately aborting the dive. Swiftly, they dropped the VBT weights and the surfacing weight and headed to the surface—over an hour and a half away.

Back on deck, they breathed a sigh of relief before delving into what had happened. "While I don't think we were in danger of having an actual fire, it was a scary moment," Vescovo said. "It's those first seconds where you think, 'holy shit,' before you return to your training and work the problem."

After the biggest electronic gremlin to date, Blades set about trying to figure out what had happened. After several days of troubleshooting and consulting with other experts in submersible electronics, Blades determined that when the manipulator arm was powered on, a current was unleashed that caused a surge to escape through the CTD, an instrument used to measure the conductivity, temperature, and pressure of seawater, and shorted it out. The current became so high that the accessory circuit breaker detected the surge and tripped the circuit, which stopped the runaway current and shut down the manipulator arm. Because the current surged, it burned out an in-cabin connector to the manipulator, thus causing the puff of smoke.

Despite this event, Vescovo was mostly pleased with the progress at Advanced Sea Trials. He emailed the team a riff on the Dos Equis commercials starring "the most interesting man in the world," the bearded character in the TV ads that bore a resemblance to him. On his dive with Lahey to 4,970 meters, Lahey videoed Victor holding a Dos Equis and delivering a variation on the ad's tag line: "I don't normally dive to the bottom of the ocean, but when I do, I prefer to take a Dos Equis."

Based on the Advanced Sea Trials dives, Struwe issued a DNV-GL class certification for the *Limiting Factor* on August 13. Struwe classed the submersible to 5,000 meters with a technical capability of full ocean depth. He recommended it not be launched in sea state 4 or greater. The *Limiting Factor* was now the first commercially certified, privately owned submersible classed to 5,000 meters.

"Very well done by all involved," Vescovo emailed the team.

The news was a boost to the morale of everyone involved heading into the first open water ocean dive at the *Titanic*. Still, the fact of the matter remained that the fate of the Five Deeps Expedition ultimately hinged on the *Limiting Factor* being fully functional and launched, diving and recovered on a continuous basis in much rougher seas.

CHAPTER 9

CIRCLING THE *TITANIC*

Titanic. It's the iconic underwater image that evokes feelings of grandeur, tragedy, mystery, and Leonardo DiCaprio romancing Kate Winslet. While diving the *Titanic* wreck would be the first chance to test the *Limiting Factor* in open ocean waters, it was really about a splashy kickoff for the Five Deeps Expedition, as this would be the first manned dive to the wreck in thirteen years. If you were going to do an open ocean sea trial, Vescovo figured, why *not* do it at the *Titanic* if you could?

In addition to serving as a launching point, successful dives at the *Titanic* would be a boost to all involved. Atlantic Productions had the most on the line. Geffen had sold the rights to a one-hour program on the *Titanic* dive to National Geographic Channel and other broadcasters. He had also arranged a media day in New York for October 19 aboard the *Pressure Drop*, which was scheduled to sail south and port there after the *Titanic* dives, to announce the start of the Five Deeps Expedition.

Triton was organizing a smaller media event the same day for the ship community. The company held the exclusive rights to sell the ship and the sub through January 31, 2019, at a minimum price of $48 million, so it needed to begin to drum up interest from possible buyers. What better way than having its first full ocean depth sub dive the *Titanic* en route to the ocean dives?

For Rob McCallum and EYOS, it was another chance to show they were the go-to company for complex expeditions.

Having the *Limiting Factor* dive the wreck of the *Titanic* would connect people to the expedition in a way that starting out at Molloy Deep or the Puerto Rico Trench would not, because it was far more relatable. "You could go to the bottom of the oceans, find a unicorn, the lost city of Atlantis, and a spaceship, and people will more likely remark: 'He went to the *Titanic*!'" McCallum said.

The RMS *Titanic* famously hit an iceberg and sank in 1912 about 370 miles southeast of Newfoundland at the nautical coordinates of 41.7° North, 49.9° West. The wreck lies at approximately 3,900 meters, or 2.4 miles below the surface, in two main pieces roughly one-third of a mile apart. Though more than 140 people had been on dives to the wreck since it was discovered by Robert Ballard in 1985, no manned submersible had been there since James Cameron and his team dived it in 2005.

Although the *Titanic* was a British ship that sank in international waters, and the Five Deeps Expedition was flying under a Marshall Islands flag, a U.S. federal court order was required for the dive. To help protect the wreck and prevent the plundering of artifacts, the U.S. government had placed the wreck under the governance of the U.S. District Court in Norfolk, Virginia. The Consolidated Appropriations Act of 2017 protects the wreck from being disturbed—by American citizens, anyway—and further bills passed by the U.S. Congress make removing artifacts from the wreck a criminal act. Only one company, RMS Titanic, Inc. (RMST), was authorized to remove and preserve artifacts from the wreck. From 1993 to 2000, the company conducted numerous dives and recovered more than 4,000 items, many of which were on display at their for-profit exhibit at the Luxor in Las Vegas.

Securing the permit was something of a bureaucratic nightmare. McCallum sent sixty-eight emails in relation to the application. The week before the expedition was to set sail for the wreck, three representatives from NOAA, one from RMST, and an assistant U.S. attorney were in court for a hearing on the application. They were all in support, but Judge Rebecca Beech Smith took four more days before signing off on the order, which prohibited removing any items or making contact with the wreck or even the seafloor around it.

The *Pressure Drop* put into St. John's Harbour in Newfoundland to pick up the Triton team, the Atlantic film crew, and several *Titanic* aficionados. Parks Stephenson, who had been on multiple *Titanic* expeditions with James Cameron, had been invited on the trip for his historical knowledge of the ship. "For me, *Titanic* still has stories to tell, and this is chance to uncover more of them," he said.

P. H. Nargeolet, Vescovo's technical adviser during the building of the *Limiting Factor* who had been on thirty dives to the wreck, was there for his expertise on both the wreck and the submersible. Microbiologist Lori Johnston, who had left the last metal plate at the wreck to better estimate its rate of decay, was on the trip to supervise placing a new tray on the wreck. Sindbad Rumney-Guggenheim, the great great grandson of Benjamin Guggenheim, a scion of the wealthy American family who perished when the ship sank, was also on board.

As the *Pressure Drop* sailed from St. John's Harbour on a sunny but chilly day in the late afternoon of September 30, everyone gathered in the sky bar, the upper most part of the ship, to take in the views of the colorful wood clapboard houses and the ancient stone battle stations once used to protect the Canadian harbor. The excitement factor was high, and there seemed to be a bit of DiCaprio's enthusiastic Jack Dawson in everyone. Vescovo had even brought a framed film poster of James Cameron's *Titanic* to hang along with the other sub-themed movie posters decorating the galley. Being slightly superstitious, the crew didn't want to hang it until after the mission was successfully completed.

It would take a day and half to sail the 330 miles to the *Titanic* site, so the first dive was planned for Tuesday morning. Though the schedule called for five dives to film around the entire wreck over eight days, the weather appeared to be conspiring against those plans.

Some 100 miles south of the *Titanic* site, a slow-moving tropical cyclone had become the twelfth named storm and the sixth hurricane of the season. Hurricane Leslie had taken shape on September 22 and meandered around the North Atlantic Ocean. Leslie fell in and out of hurricane status, but regained the distinction on September 28 when it merged with a second frontal system. Projections indicated that Leslie could pass near the *Titanic* site on Friday or Saturday.

The plan was to do as much as possible on the first dive in case the weather was too bad for additional dives. The primary goals were to collect an authorized rusticle, slivers of rust that grew off the iron sides of the *Titanic* from bacterial activity, and also recover and replace a tray placed on the wreck that measured its rate of deterioration.

"We're taking a big roll of the dice because we are a victim of the weather," Vescovo said.

❧

As this was the first extended voyage of the expedition, it would be the longest the team had lived and worked together on the ship. It didn't take long for a rhythm of life on the ship to be established. A morning meeting was set for after breakfast, and then each team went off to prepare for what lay ahead.

On the first afternoon, the entire ship gathered for a ten-minute safety briefing by the first officer. As everyone put their life jackets on, Lahey joked, "Please tell me we have more lifeboats than the *Titanic*."

Everyone ate three meals together. Typically, the ship's crew ate first and fast. For the others, meals became work and social time. The center of most conversations was what could be seen at the *Titanic* wreck.

The quality of the food and the decision to allow alcohol were two critical aspects to the morale of any expedition. Initially, Vescovo was going to hold to the U.S. Navy standards of no alcohol on board, but McCallum, a veteran of long journeys at sea, convinced him otherwise. "You should never go out of sight of land without alcohol," he said. The compromise was British maritime rules: beer and wine, though the occasional bottle of Crown Royal popped up.

Each afternoon, three large plastic coolers on the upper deck were stocked with wine and beer. Dubbed the sky bar, the area had large, white plastic picnic tables and three stationary deck chairs facing the bow of the ship. After dinner, even in some of the coldest weather, people would drift up to the sky bar to unwind.

The galley was run by chef Manfred Umfahrer, an Austrian who has a friendly round face and a goatee with a point that dangles two inches from his face. Like the other members of the hotel staff, he had lived his life on ships, and didn't seem to mind the long hours and endless days at sea.

Umfahrer rose each day at 4:00 A.M. to begin prepping breakfast. Set meal times were posted. Though breakfast was scheduled for 6:30 A.M., the room was full of crew by 6:15 and empty by 6:45. Lunch was 11:30 A.M. to 12:30 P.M. and dinner from 5:00 to 6:30 P.M.

The ship can carry enough food for up to one and a half months at sea at full capacity of forty-nine. The mess level has a dry storage room with metal shelves containing vats of olives, peppers, and oatmeal. The walk-in refrigerator has large baskets with 100 pieces of fruit in each one. The freezer is stocked with twenty-five pound-roasts and dozens of whole fish, often purchased from local fishermen when the ship ported.

In the mess room, the L-shaped buffet features a cold side and hot side. Above each hot dish is a small, black chalk board with the name of the dish written in cursive on it. Four entrées are the minimum, six are more like it. In deference to the predominantly Filipino crew, every meal has one or two Asian dishes, such as fried rice or spring rolls.

The one certainty is flavor and spice. Tacos are accompanied by three salsas, hot, hotter, and hottest. Enchiladas are stuffed with chopped jalapeños. Burritos have a kick that makes them taste like authentic Mexican street food.

"The only thing more important than good food is a working coffee machine," McCallum says.

The biggest complaint would become a recurring one: the Wi-Fi was inconsistent and slow when it did work, despite considerable investments to have up to 5 megabits per second download speed ship-wide. The issue was that the bandwidth was divided into groups and when Vescovo was online, more than half of it was automatically allocated to him. Though he had begun to pare down his business commitments in preparation for the deep ocean dives, reducing his load from five to four major companies, he still needed to be in constant contact with his office. But as much as people complained about the poor Wi-Fi, no one had an issue with the owner taking up so much bandwidth.

"Someone has to pay the bills," McCallum said with a wink.

~

After dinner on the first night at sea, Geffen prepared a press release to send out after the impending first dive that was cleared by National Geographic.

AT THE SITE OF THE TITANIC WRECK, NORTH ATLANTIC OCEAN—An expedition led by Caladan Oceanic today announces that they have completed the first manned submersible dives to the Titanic wreck in 13 years. Following established U.S. legal protocols, under NOAA's oversight, a team of experts and scientists have been examining the remains of the ship to assess the current condition and project its future. The findings of this expedition will air as a special on the National Geographic Channel and other networks globally in December 2018.

But by the next morning, it began to look like the word "almost" would need to be inserted. When the ship reached the *Titanic* site, the weather was worse than expected. Swells were coming from two directions. Hurricane Leslie was now a massive Category 1 hurricane the size of Connecticut, and though slow moving, it was creeping toward the site from the south at the same time a separate weather front was bearing down from Canada to the north. To complete the trifecta, strong currents were coming across from the European side.

"There's no accurate weather modeling this far out," Captain Buckle explained. "At home the weathermen from town to town can see the weather and report on to the next guy and so forth. In the old days, ships used to wait to feel the swells to know where a storm or hurricane is coming from. Now there are satellite models, but out here they are still best guesses."

The ship was constantly pitching and yawing, left to right, left to right, causing it to feel like you were on an uneven rollercoaster moving in slow motion. Walking required spreading your legs for stability, the "seaman's walk." Sea sickness tablets were a must for all but the most seasoned. On the trip, more than one person would bump their head on the beams inside the staircase, and the occasional glass would slide off a table.

That Tuesday afternoon, a memorial service was held at the site where the *Titanic* sank. With the cameras rolling, McCallum spoke about the tradition of honoring the victims and tossed a wreath overboard. It was the day's only real activity.

By Wednesday morning, the winds were kicking up as high as 43 knots (50 mph), and the ship was rocking badly. There was no realistic chance to attempt a dive. Lahey, Vescovo, McCallum, Geffen, and Buckle all began discussing what to do. Returning to St. John's and waiting out the storm wasn't an option because of the transit time and the fact that the ship needed to be in New York for the upcoming media day.

Geffen was particularly concerned because the trip was eating up his entire budget for the *Titanic* film, so he was pushing to stay as long as safely possible. If he wasn't able to film at the wreck, he would have to wait another year when the ship headed to the Molloy Deep, as well as having to fund that shoot out of his own pocket. Vescovo wanted to get the footage now, if at all possible, so he didn't have to spend the money to return to the site.

A decision was made to wait one more day, as conditions looked slightly more favorable on Thursday. Buckle had already concluded that Friday was out, because the ship needed time to sail inland in the event Hurricane Leslie continued on its current path.

On Thursday morning, Geffen, McCallum, Lahey, and Buckle were on deck at 6:00 A.M. surveying the conditions. A misty rain was falling, but the sea had calmed slightly and winds were down to 10 knots. They decided to monitor the conditions over the next two hours to determine if a short dive would be prudent.

Lahey and Triton's Magee were against diving. "An eight-hour window is not enough to dive," Magee said. "If there was an issue with the sub on the bottom, say it became entrapped in the wreck, it has ninety-six hours of oxygen and we have the assets to go and get it, but not if the ship is blown off the site by a hurricane."

Lahey shook his head. "That would not be a pretty picture."

Still, Geffen was slightly encouraged by the calmer seas. He met with his production team and made final plans to film the dive.

But at 9:00 A.M., the dive was called off by Vescovo. The weather was holding, but it was determined that the window was too narrow and it was just too dangerous.

"The weather is the limiting factor, no pun intended," Lahey said.

Vescovo added, "I feel like the Dallas Cowboys . . . better luck next year."

∞

In a last-ditch effort so as not to have entirely wasted the time and cost of steaming up from Florida and out to the *Titanic* wreck, a revised plan was made. The *Pressure Drop* would sail toward St. John's and the team would attempt an open water dive on Friday closer on the Grand Banks of Newfoundland, a massive underwater plateau off the coast of Canada, where the weather looked to be more favorable. The idea was to practice the launch and recovery of the sub and continue to test its performance.

When the ship arrived at the targeted area, the sea state was elevated to between a 2 and 3, but conditions were far better than at the *Titanic* site. Vescovo, McCallum, Buckle, and Lahey met early Friday morning to discuss whether to dive. Lahey didn't want to risk damaging the sub by attempting to launch and recover in more strenuous conditions than his team had dealt with to date. Buckle felt the dive was doable.

McCallum spoke up, stating the obvious. "We need to show ourselves that we can either do this or we can't," he said. "And if we definitely can't, we might as well pack up and go home because there's no point of heading off to the Southern Ocean."

Lahey shook his head. "There's no fucking way we should dive in this," he said. "It's an unnecessary risk. We'll be diving in Puerto Rico and we can work on the dynamic recovery there."

The Triton crew had never attempted a launch and recovery in conditions bordering on sea state 3. In fact, the toughest seas they had faced were at Advanced Sea Trials, which was just over a sea state 1. Based on the track record to date in the Bahamas, there was little to suggest the dive would be successful in these conditions.

Vescovo listened to all of the voices of the people he would be relying on for guidance over the coming year. He concluded that McCallum was right, they needed to try, because this was the last chance to test themselves in big waters before the potentially violent seas of the Southern Ocean. He had to know if the team could perform.

On the white board that listed the day's plan, McCallum wrote: "Full dress rehearsal for Puerto Rico."

Two hours later, the LARS team readied the *Limiting Factor.* Carrying a pre-dive checklist, Vescovo walked around the sub with Lahey, checking off items one by one. Once they were satisfied, the *Limiting Factor* was attached to the A-frame crane by the main line and moved along the track into position at the back of the aft deck.

Cameras were rolling everywhere. A cameraman was on the *Xeno.* Another camera was filming the deck operations. A drone was flying above, and a diver with a helmet cam was in the water to film the sub from below as it began its descent.

The wild cards were the cameras attached to the sub. Atlantic had rented the cameras from the Woods Hole Oceanographic Institution and contracted for two WHOI personnel to be onboard in the event of problems. The cameras had been delivered to the ship in Newfoundland, leaving no time to test them. Triton had to make significant, if not downright unsightly, modifications to the sub to accommodate the cameras, and it wasn't known if the cameras would affect the sub's electronic functions in the water.

Just before the dive conditions leveled off, though the swells were still creating white caps. The sky was shrouded in a hazy fog, and a steady wind was blowing. The *Learned Response* support boat that had drawn criticism from Buckle was launched without an issue. The *Xeno,* which would take Vescovo and Lahey to the sub, was next. The *Xeno* circled around to pick up them up. As Vescovo was climbing into the *Xeno,* he slipped off of the ladder while transitioning to the boat and fell backward onto the gas tank and bruised a rib. He quickly shook it off.

Kelvin Magee was directing the launch from the bottom deck and Frank Lombardo was operating the A-frame controls. Lombardo, an experienced deck hand with subs, lowered the A-frame as far away from the ship as it would extend, only about five feet. This was the issue that had been identified when the ship was purchased, that the A-frame was not long enough to lower the sub a comfortable distance from the ship. The twelve-ton sub dangled from the A-frame, twisting in the wind.

"Down on the main," Magee said, making a fist and twisting it downward.

Lombardo moved the control arms and slowly lowered the sub. When it was even with the stern, a gust of wind whipped up and slammed the sub into the ship. A loud crunching noise was heard when the sub made contact with the stern.

"Fuck!" Lombardo said, as Magee yelled: "Up on the main! Up on main!"

Lombardo yanked on the up lever, and the sub was lifted above the ship.

A slight crack in the white syntactic foam was visible.

Magee went inspected. "Looks cosmetic," Magee shouted. "Let's keep going . . . down on the main . . ."

A visibly frustrated Lombardo lowered the sub again, this time successfully, into the water. Though the sub was just a few feet from the ship, the drag from a sea anchor pulled it away from the stern as the ship was moving through the sea at a slow speed to maintain tension on it. The sub twisted left and right as the crosscurrent hit it.

The *Xeno* approached the sub, but the driver was afraid to get too close for fear of smashing into the cameras mounted on it. This caused Vescovo and Lahey to struggle when grabbing the handrail to climb onto the sub. After several attempts, they managed to board the sub.

Once Vescovo and Lahey were inside the *Limiting Factor* "the swimmer," Steve Chappell, closed the hatch. He then went to work, first removing the railing from the sub and then detaching the towline from the ship. The process took nearly fifteen minutes as the sub tossed like a bucking bronco. When he was finished, he dove into the water and swam to the *Xeno*.

"Are we released?" Vescovo asked the control room.

"Roger that . . ." came the reply.

After final checks and clearance given to dive, the sub slowly disappeared from sight.

The pilots checked in every fifteen minutes, giving their depth and affirming "life support good." Communications during the dive were audible but garbled. Vescovo and Lahey attempted to navigate to the lander dropped earlier but didn't find it. Ninety minutes into the welcome, uneventful dive, the *Limiting Factor* asked for clearance to surface.

For the recovery, Buckle was able to maneuver the ship into a position that blocked the Labrador Current coming down from Canada, the culprit of the choppy seas. This would give the sub more time to pump the water out of the main ballast tanks in the rough conditions. Still, the recovery took far too long for both comfort and safety's sake.

Because of the rough seas, the swimmer Chappell had difficulty attaching the tow line and the sea anchor. Once the sub was in position to be pulled onto the ship, he struggled to clear the water inside of the trunking using the manual pump before he opened the hatch. After ten minutes, he resorted to bath towels. For such a high-tech sub, it was a decidedly low-tech solution that garnered snickers from the onlookers on the ship.

Back on deck, Lahey said, "The sub was about to become a vomitorium."

Multiple problems had occurred inside the sub during the dive. The worst was a small hatch leak. There was enough water ingress that by the end of the ninety-minute dive there was a small puddle of standing water on the bottom of the sub capsule. Though not a safety hazard on a short dive, it would cause a longer dive to be aborted.

The VBT weights jammed on both the starboard and port sides after the first few had released. On all eleven of the sub's dives, the VBT weights had jammed on one side or the other at some point. More work was needed there.

On the electrical front, the pilot's GUI had intermittent failures that caused it to freeze, necessitating rebooting the monitor four times during the dive. The foreground camera used by the pilot for navigation was inoperable.

Even worse than the nonfunctioning foreground camera, one of the exterior cameras provided by WHOI did not work at all, and the other produced substandard footage. Given that Atlantic was looking in the region of $100,000 for the *Titanic* expedition in camera costs for these, Geffen was understandably displeased. He and Syder, his head of production, concluded they would have to look elsewhere for more reliable cameras. However, they were also relieved that the dive hadn't been the lone one to the *Titanic*.

That night after dinner, many of the Triton team gathered in the lounge. Several bottles of red wine and a bottle of Crown Royal were on the table in the middle of the room. They sat around drinking and reviewing the day's events.

"Gentlemen, we got our asses kicked today," Lahey declared. "But we'll get it right."

Sipping a beer in the sky bar, McCallum put a positive spin on the day's events. "We got a real fright, but it was a fright that we needed," he said. "This was the first time we showed we could operate in sea states greater than a meter. We need to break things and make mistakes before we hit harsh conditions. These will be elements on the road to success."

Despite not being able to trumpet a dive to the *Titanic*, the media launch was still on for October 19 in New York. The *Pressure Drop* would sail from Newfoundland to Staten Island. Discovery Channel executives and key media would be invited onboard for a tour and a presentation of the expedition given by the team leaders. The hope was that the media seeing the *Limiting Factor* and footage of it in action would generate some buzz about the Five Deeps Expedition.

Following the media event, the sub needed to be taken back to Triton's new facility in Sebastian, Florida, and completely disassembled to ascertain and fix the cause of the hatch leak, as well as the other items that had failed.

"You have my personal guarantee, we will not see any water in the sphere through the mating ring again," Lahey said to Vescovo.

The following morning, the *Pressure Drop* pulled into St. John's Harbour. Everyone who was not part of the transit crew disembarked. The framed *Titanic* film poster lay unhung on its side leaning against a cabinet in the lounge, a metaphor for the sideways journey that this leg of the expedition had become.

CHAPTER 10

WEATHERING STORMS BIG AND SMALL

Time was running short for Captain Stuart Buckle to leave port in Newfoundland to arrive in New York in time for the planned media launch of the Five Deeps Expedition. Hurricane Leslie, which had knocked the *Pressure Drop* off the *Titanic* site, was still churning in the North Atlantic, just above Newfoundland. Compounding the problem were the remnants of Hurricane Michael, which had come up the east coast of the U.S., skirted Nova Scotia and Newfoundland, and was finally heading out into the open waters.

Buckle waited until the worst of Hurricane Michael had passed. On October 14, with winds still gusting to 68 knots and a twenty-foot sea swell, Buckle decided that conditions were safe enough to sail for New York, giving him four days to make the October 19 event. The captain of a 180-meter ship holed up in St. John's Harbour radioed and called him a "crazy little guy" for sailing in such conditions.

As the *Pressure Drop* hit the open waters, high winds and sea swells slowed the ship from its average of 9–10 knots to 6–7 knots, pushing the estimated arrival time from midday on the 18th to the morning of the 19th, the day of the event. The following morning, it became clear that the ship would not make it in time for the afternoon event.

Geffen discussed the situation with Discovery Channel executives and they decided to proceed and hold the event in a conference room at its midtown offices. The fact that the launch would be missing its star—the submersible—added to the existing tension among the principals. In the two weeks preceding the event, an email skirmish had broken out over the media launch and who could do what regarding press.

Though Atlantic Productions had contractual control on behalf of Discovery over the expedition's media as part of its deal with Vescovo, Triton was pushing back in an effort to protect its marketing interests. Both Bruce Jones and Lahey claimed they were not bound by Vescovo's contract with Atlantic as they weren't party to it, and so they had scheduled their own event aboard the ship. They held the exclusive rights until February 1 to sell the ship and the sub, and they felt they needed their own press coverage to reach interested buyers.

Geffen was among the best producers in the nonfiction film world at promoting his projects, but he told Vescovo that Discovery was insisting on holding back the release of more information to generate the maximum amount of buzz all at once. He also had considerable experience with multilayered projects. Discovery's position, he explained to Vescovo, was that Triton should invite its media list to the main event, and not do anything separate in an uncoordinated manner.

But while Caladan and Atlantic had agreed that the media strategy would be jointly coordinated, Triton felt this was limiting their ability to market the sub.

A compromise was reached allowing Triton to hold a smaller event on the ship with industry press. However, when it was evident that the ship would not make it to New York, Triton canceled its event and invited their press contacts to the Discovery event.

The day before the media launch, the team leaders all attended a small event at the Explorers Club. Vescovo had recently become a member of the Explorers Club and had completed its "Grand Slam," summiting the highest peaks on each of the seven continents and skiing at least 100 kilometers at the North and South poles. With the planned deep dives, he was now taking things to a whole new level. At the club, he was presented with a flag to take with him in the sub to the bottom of the five oceans. When the expedition was complete, the flag would take its place among the others hanging in the club, including one from the first Moon landing.

That night, the group gathered for dinner at Sip Sak, a Turkish restaurant in midtown Manhattan. Lahey, Jamieson, McCallum, and P. H. Nargeolet were all there. Vescovo arrived a half hour late, after doing a radio interview that ran over. Geffen had other plans and did not attend the dinner.

After a spirited dinner, Vescovo invited everyone to see his apartment in Trump World Tower by the United Nations. On the walk over, Lahey quipped to McCallum, "Will the Donald be there?" Someone who overheard the comment pointed out that it was the first mention of Trump he had heard in all the hours the team leaders had been together.

The mood was light, thanks to heavy pours of Turkish wine from the restaurant's extremely friendly proprietor. The group rode the elevators to the 76th-floor apartment, which has one of those signature New York high-rise vistas in which the building lights seem to stretch forever. The apartment was decorated with touches of Vescovo's love of exploring, and the refrigerator, as he self-depreciatingly pointed out, was filled only with cans of Diet Coke, his daily fuel.

As a social evening away from the ship, it went well. There was a feeling of unity and that everyone was in this together. As Lahey summed it up, "Pretty fuckin' fun night."

∞

The following day was a contrast in possibility and reality. The expedition team gathered in a conference room at Discovery's offices to announce a five-part series on the Five Deeps Expedition and to discuss deep ocean exploration. The first press had appeared that morning, a small piece in the TV section of the *New York Post*, an exclusive given to the paper. The expedition's website—fivedeeps.com—had just gone live.

Discovery had invited several press outlets to the event, including the BBC, Reuters, the Associated Press, and CNN. The Richards Group, a Dallas-based PR and marketing firm that Vescovo had hired to represent him, invited the *Wall Street Journal*'s science editor, a *Forbes* writer, three producers for CBS's *Sunday Morning*, and an *ABC World News* correspondent. However, because the sub didn't make it, none of the high-end organizations attended.

But the show went on anyway. Geffen moderated a panel discussion with Vescovo, Lahey, Ramsay, Jamieson, and McCallum, and showed clips of the sub diving. There was no shortage of platitudes and hype.

Geffen opened by talking about Atlantic's 25-year working relationship with Discovery. "We've done some pretty big things, but I think they realize this could top most of them to follow this expedition," he said.

Lahey called the sub "the equivalent of a moonshot." Geffen introduced Ramsay, the designer, as "the new Jony Ive," a reference to the legendary head of product design at Apple, and he referred to McCallum as a "mastermind" in planning. McCallum declared that the sub was "the most significant exploration vehicle since Apollo 11." Vescovo was always embarrassed by those who made this—to him—very hyperbolic comparison. On the science side, Jamieson talked about how the ocean could only be fully understood by studying its depths, a door that the *Limiting Factor* was now set to open, adding, "this could be a groundbreaking expedition in terms of understanding the Hadal Zone."

However, at one point, there was an awkward exchange. When Geffen was talking about diving to 2,000 meters in one of Ray Dalio's subs that had been built by Triton, Ramsay perked up. "I do hope you didn't take one of our other subs to 2,000 meters because it's 1,000-meter rated," Ramsay said.

"We definitely went to at least 1,600 meters," Geffen countered.

Lahey stepped into the breach. "You wouldn't have gone below 1,000 meters," he said. "It could be 1,600 feet."

"No . . . it was 1,600 meters," Geffen said, before changing the subject.

At the end, McCallum summed up the expedition in his irreverent manner. "We're taking a group of people who haven't worked together before and we are sending them off on a mission that has never been done before in a piece of machinery that has never operated before," he said. "What could possibly go wrong?"

As everyone on the dais knew, plenty could and already had gone wrong. The reality of the situation was slightly different than the optimism filling the presentation. At present, instead of being docked in Staten Island for media tours, the *Pressure Drop* was fifty miles off the coast of Virginia, caked with salt from the battering it took dodging two hurricanes, carrying a crew that was exhausted from the ordeal. The ship was in the process of transporting the star of the series, the *Limiting Factor*, back to Triton's facility in Florida for refit.

Because there was no sub to see, the resulting press coverage was tepid. There were no TV cameras at the event. The BBC and Sky News did features on the expedition after conducting one-on-one interviews with Vescovo and Lahey. One science reporter from Reuters bizarrely asked Vescovo in an interview how he would get in and out of the sub—at the bottom of the ocean. Vescovo politely explained how that would not be possible. *Forbes* posted a story online based on information it was provided, as its reporter did not attend. However, the story contained multiple inaccuracies—it even misstated the name of the sub—and accusations began flying back and forth over who was responsible.

The biggest problem with the *Forbes* story was a quote attributed to Geffen, which Geffen said did not come from him. He was quoted saying that the ocean "can be used for medical innovations such as finding cures for Alzheimer's." Since 2010, researchers had been studying whether the sea squirt, human's closest invertebrate relative, could aid in Alzheimer's drug development by testing whether the accumulation of plaques and tangles that mar the brain could be undone in sea squirts. But this was far beyond the scope of any science the expedition was contemplating and, in fact, could create regulatory problems with countries whose shores bordered the oceans.

Jamieson explained that merely saying the expedition was looking for pharmaceuticals or cures for cancers or Alzheimer's could endanger their permits. The reason was that anything in a country's Exclusive Economic Zone (EEZ), the area stretching 200 nautical miles from its shore, is the property of that nation under the 1982 United Nations Convention on the Law of the Sea.

"We cannot simply go into someone else's EEZ and go looking for potential pharma, without permission," the scientist emailed Vescovo. "That is THEFT. I led the expedition 'PharmaDEEP' to Antarctica in 2015, where we did have permits, and 6 months later I was nearly ARRESTED for breaking the agreement as the person who had clearance to allow screening for pharma and passed the sample on to someone who had not. I didn't even know. In the event of 'illegal' prospecting it is the COLLECTOR who goes to prison, and in PharmaDEEP and indeed the 5-deeps, that is me. I ain't going to jail on this one!"

Vescovo's position was unequivocal: the expedition needed to make sure it was clear to the press that there was no biomedical prospecting occurring. If there was any possibility of the dive permits being jeopardized, Vescovo said that he would pull the science mission from the expedition from that leg, or *in toto*, to ensure the permits needed to dive remained in place. To prevent a potential conflict, there needed to be better—and accurate—communication with the media.

For his part, McCallum thought the New York event was a poorly produced waste of time. He wrote Vescovo an email calling the media day "amateur night" and criticized Geffen for grandstanding about past Atlantic films, for speaking with cue cards and for the quote in *Forbes*. He concluded that the "riff-raff media will always get it a bit wrong, but *Forbes* is a pretty good player so I think they were given the wrong info."

For his part, Geffen pointed out that a Discovery executive was supposed to be the MC for the event, but Geffen had to step in last minute. And of course, the media event had been planned around the sub being there—and when it wasn't, then obviously the event was less exciting and drew fewer journalists.

In the rush of responding to the over 100 emails a day he was fielding, Vescovo accidentally emailed McCallum's critique to Geffen as part of a forwarded email chain about correcting the *Forbes* story.

"Anthony was upset and called me," Vescovo said. "He said, 'Rob is unprofessional, but I will work with him.' For all Rob's gripes about Anthony, Anthony has equal gripes about Rob and the way he handles the expedition."

Vescovo was frustrated at the tense relationship that was developing between the two strong personalities before the expedition had even gotten underway. He himself was not particularly pleased with certain aspects of the media coverage thus far, but also didn't think McCallum appreciated that Geffen had delivered a five-part Discovery Channel series and a multi-broadcast coproduction (including National Geographic Channel) on the *Titanic*. "These are no mean feats that I don't think Rob fully appreciates," he said. "I do, anyway."

Geffen told Vescovo that he did not speak directly to the *Forbes* reporter and that she had taken the quotation out of context from press materials that Discovery sent out, as one of the scientists who planned to join the expedition in the Southern Ocean had done research in that area. As for the debate over the depth he dived in Dalio's Triton sub, Geffen brushed it off as "banter by a Brit confusing meters and feet."

For whatever reason, perhaps because of all the back and forth or in spite of it, the *Forbes* story was never corrected.

Small skirmishes like this one were causing Vescovo to become increasingly frustrated with the overall state of affairs. For all that had gone wrong in private, announcing the Five Deeps Expedition publicly would mean that any future failures would be out in the open—widely so. For starters, it wasn't clear when—or even if—the first dive would actually come off.

The *Limiting Factor*, which had only been partially functioning as it should, was being completely disassembled for repairs, and the *Pressure Drop* was headed to dry dock to be outfitted with a massive new sonar. These events needed to come to a successful conclusion before the Puerto Rico Trench dive could happen in mid-December, or the dive would have to be postponed so that expedition could make the mid-January weather window in the Southern Ocean.

"At the end of the day, it was Rob's recommendation to try for *Titanic*, as well as saying we could make New York," Vescovo said. "I know, it was weather that held us up, but at the end of the day, we didn't deliver, and we had a poor schedule to make both happen. He didn't put enough safety margin in, and I ended up wasting maybe $300,000, at least, steaming up to St. John's and now back down on a failed mission. And again, I know he would scream 'weather! hurricanes!' but even those need to be factored into planning and if we have any major issues on the PR Trench, I'm going to have to start asking hard questions of Rob as well as Triton on why we keep failing in missions."

McCallum took it all in stride. "That's why I had a name tag printed for myself with 'Scapegoat' on it," he said. "My job is to take the criticism and make things work."

&

The *Pressure Drop* arrived in Fort Pierce, Florida, on October 25 and offloaded the *Limiting Factor* for its refit at Triton's new, larger facility in nearby Sebastian. The ship then headed for dry dock in Curaçao, off the coast of Venezuela, where it would be outfitted with the sonar. Because of the tight schedule, the *Limiting Factor* would need to be transported to Puerto Rico to rejoin the ship there for the first planned deep dive to the Puerto Rico Trench, the deepest part of the Atlantic Ocean. The situation was less than optimal, and a bit risky, but it was the only way to fix the sub and install the sonar at the same time.

The day the sub arrived, Triton sent an estimate of costs to Dick DeShazo for the maintenance, totaling $146,846. This included fixing the sphere leak, the VBT weight system, the air conditioning system, the trunk pump, as well as multiple electronic issues. "I gulped when I saw it," DeShazo said.

DeShazo forwarded the estimate to Vescovo, who was understandably miffed. He felt that many of the items were design flaws and therefore should be covered as warranty items, and he asked DeShazo to detail them. Still, he said that he was okay funding the estimate "for the sake of domestic peace," though he wanted the warranty items "where there is clear poor workmanship or a screw-up by Triton" applied to Triton's side of the ledger in the onward sale agreement.

"They have gotten into the awful habit of just throwing rather large costs, willy-nilly, into a bucket that charged Caladan and expecting me to pay 100 percent, on time, with no explanations, cost backup, or pushback," Vescovo said. "It's getting out of hand and as a financially challenged organization, they need to do things far better when it comes to cost estimating, taking responsibility for their errors, and billing properly, not to mention being far more sensitive to cost overruns and managing costs."

Sub designer John Ramsay quickly diagnosed the biggest problem, the leak, and came up with a fix. He found that the leaking in the pressure hull was the result of the opposing halves of the main pressure hull sliding, ever so slightly relative to each other against the equatorial ring plate around them, whenever the sub was lifted by crane. This was being caused by movement in the frame attached to the hull that held the syntactic foam. His solution, which he cleared with certifier DNV-GL, was to install what he called a Circumferential Preload System. This basically entailed placing a tight metal band around the hull at an angle to prevent the two halves from shifting.

"I knew right away what the issue was, but it took a while to come up with a sure-thing fix," Ramsay said. "I'm pretty sure it won't be leaking anymore."

~

While the *Limiting Factor* was undergoing its repairs, the *Pressure Drop* headed for the Damen shipyard in Curaçao. Once out of the water, a gondola on the keel would be installed, followed by the massive sonar. Damen was also contracted to build out a new davit, the small crane used to lower the *Learned Response* support boat, as a solution to replacing the boat altogether.

Released in 2018, the EM 124 sonar was the fifth generation of a range of new multibeam systems from Kongsberg Maritime, a Norwegian company. It is the most advanced civilian multibeam sonar echo sounder available and could generate high-resolution seabed images for mapping from shallow waters to full ocean depths with quite wide "swath" coverage and resolution. The sonar would allow the expedition to map the ocean trenches and to determine with great accuracy the true depths of the oceans. This was the first one Kongsberg had sold; thus, it carried the serial number 001.

But before the installation of the gondola and sonar even began, there were two major vendor failures and a price increase to boot. Forty-eight hours after DeShazo accepted Damen's bid on the advice of Thome Croatia, the company hired by Vescovo's Caladan to oversee the ship's operations, and wired $100,000 to start the project, Damen advised him of a 15 percent jump in price over the agreed-upon quote. Next, Damen notified him that there would be a delay of several weeks, as it had ordered the wrong type of steel for the gondola.

Then, Kongsberg literally misplaced its $1 million sonar during the shipping process. The sonar was to be loaded onto a cargo ship in Norway. The Kongsberg team in Curaçao waiting for the sonar tracked the ship and its arrival—but it turned out that the sonar had not been loaded on the ship and was still in Norway. Kongsberg scrambled to send the sonar by air freight and agreed to eat the $50,000 shipping cost.

While all this was being sorted out, the *Pressure Drop* was killing time off the coast of Haiti waiting for word on when to port. "I think the authorities were a little suspicious of us and what we were doing," Buckle said. "I was glad to get word to port."

Pulling the contract from Damen was not a viable option so the price increase was reluctantly agreed to—it was still a "fair" price in Vescovo's view, who had actually worked in a shipyard at one point in his career. As a practical measure, the expedition could not afford the time to find another shipyard and still have a realistic shot at making the Southern Ocean weather window.

The ensuing weeks were disorderly. Thome had a representative on-site, but he had little success pushing Damen to move faster. Vescovo ended up hiring Brian Gamet, a sonar expert from Geosight Land & Hydrographic Surveys, to serve as the owner's rep and supervise the installation of the sonar package.

McCallum, who had initially recommended the shipyard, was caught in a tough spot. He was both pushing the Curacao shipyard to speed up the job, while simultaneously threatening to take up the matter of the inflated costs and delays with the higher-ups at Damen headquarters.

Midway through the process, Damen replaced its shipyard manager with a more experienced and responsive one, and things started to improve. Still, what was supposed to be a two-week job was turning into a six-week one. Things were not under control, and Damen was not responding as Vescovo knew they should, or could.

On November 23, an increasingly frustrated Vescovo summarily dropped what he was doing and flew in his jet with his copilot all the way to Curaçao. He and Buckle demanded a meeting with the new shipyard manager, which they were granted. The executive sheepishly admitted to a series of start-up errors and promised to accelerate the project without sacrificing quality. Vescovo was warily satisfied with the meeting and asked Buckle to maintain a close eye on the work.

With Buckle keeping close tabs, the sonar installation was finished and the ship was refloated on December 8. That afternoon, the Damen shipyard operations manager handed Buckle an invoice for $981,000, bringing the dockyard total costs to over $1.1 million, or roughly double the amount of what had been agreed to in a fixed-price contract. It had become apparent that Damen had underbid the contract to secure it, and then racked up time and materials costs, and were now threatening to hold the ship hostage until the bill was paid. This was, unfortunately, a routine occurrence in what Vescovo called "the unscrupulous world of ship repair."

Vescovo was in a no-win position in the short term. The contract carried a dispute resolution mechanism: arbitration in Rotterdam. Matt Lipton, Vescovo's attorney, found an attorney there in the event that an arbitration over costs, or the threat of one, became necessary. However, that didn't solve the current dilemma of getting the ship released.

McCallum contacted the shipyard's managing director, Lodewijk Franken, and insisted that Damen release the ship for the agreed-upon amount in the fixed price contract. "We expect Damen to honour that contract in its entirety," he emailed. "The Damen name is riding on that outcome."

Franken responded that he did not want to enter arbitration and suggested the parties settle the matter "around the table" after he prepared a justification for additional charges. He conceded that the some of the charges were incorrect, and agreed to allow the ship to sail if Buckle signed a completion certificate and Caladan paid

85 percent of the previously agreed-upon amount. DeShazo promptly wired $200,000 so the ship could sail for Puerto Rico, which it did on December 10 with an ETA of December 13, to a massive collective sigh of relief from everyone on the expedition team.

It turned out that the fixed price contract was full of holes and had not been properly vetted by the ship's management firm, Thome, whom Vescovo and Lipton had relied on to ensure it was a fair contract *without* holes. DeShazo had signed the contract and wired the deposit while riding in a taxi to the media launch event. He had done so relying on the advice of Roko Stanic, the senior manager at Thome Croatia. However, it turned out that Stanic and McCallum, who had recommended Damen, had just passed along the contract without conducting any due diligence on the Curaçao yard or performing a thorough review of the contract.

"Ultimate responsibility lies with me, and I should have noticed the lackadaisical attitude of Thome, the lower level of involvement of Rob, and trusting Dick to make sure the contract was complete," Vescovo said. "It is turning out, to my surprise, that I have a lot more general business experience than the other members of the team and I should have taken more direct control over this contract. But Jesus, I can't do everything or be everywhere, and I am paying a lot of money to people . . . to handle these details, but unfortunately, people sometimes just don't execute. They like to say they are, and tout their maritime experience, like Rob or Bruce Jones did, but when it comes to results, people just too often can fall short. I trusted the so-called experts, and got lightly burned. And good Lord, if we hadn't gone all hands on deck, and if Stu Buckle, who is just extraordinary, hadn't been there and stepped up, we would have been there another month."

DeShazo prepared a minutely detailed, 21-page charge-rebuttal memo outlining Caladan's position on every single billable item and challenging certain overages and many of the change orders. He then subtracted the $304,541 that he was disputing, and paid the remainder of the bill, bringing the total outlay to $686,560, far less than $1.081 million billed. Vescovo swore he would go to war with them, on principle, if they challenged even a single line item of their rebuttal. He never had to—Damen never responded.

~

As the work continued on the *Limiting Factor*, Vescovo, Lahey, and DeShazo explored options for safely transporting the *Limiting Factor* to Puerto Rico to rejoin the *Pressure Drop*. The three options were to ship the *LF* in its special standard container to Puerto Rico for pickup, to hire a dedicated ship to move it from Florida and do an at-sea transfer, or to find a "friend of Triton," such as Ray Dalio's ship the *Alucia*, to transport it in exchange for a dive or some other consideration.

"I feel like I have to micro-manage Triton on the *LF* shipping," Vescovo said. "I don't trust them 100 percent on this. They can be far, far too overconfident on things they don't do regularly—or even regularly—and as an organization are just overly optimistic a lot of the time. But this transport was a huge deal, and I wasn't sure they fully appreciated the stakes."

In addition to damage or loss, there were concerns with customs in Puerto Rico even though it is a U.S. territory. "The last thing we want is the *LF* sitting in a debarkation area under lock and key, waiting for paperwork or a bureaucratic snafu, or god forbid, the container being sent to the wrong port," Vescovo told DeShazo.

The dedicated ship option priced out between $100,000 and $140,000, money that could be better spent elsewhere assuming a reliable cargo company could even be engaged. Lahey estimated the third option could be in the same neighborhood, as any ship would have to relocate. The decision was somewhat reluctantly made to ship the sub on a TOTE Maritime cargo ship, at a far more reasonable cost of $25,200.

On December 5, the *Limiting Factor* was loaded onto a 40-foot, flat rack truck and secured with six lashings rated to 5,000 kilograms (11,000 pounds) each for the trip from Triton's facility to the port in Jacksonville. Kelvin Magee and a colleague followed the truck and then supervised the loading of the sub onto the cargo ship. "We had a lot of strange looks from other drivers," Magee said.

"This transport of the *LF* scares the bejeezus out of me," Vescovo said. "I have nightmares of the container getting lost, on fire, or falling overseas during the transit. I keep recalling that Jim Cameron's *Deepsea Challenger* caught on fire while it was being transported."

⁂

On December 10, right on schedule, the TOTE Maritime cargo ship carrying the *Limiting Factor* arrived in Puerto Rico and was cleared for entry by Departamento de Hacienda de Puerto Rico, the nation's Department of Treasury. Lahey and, at Vescovo's insistence, DeShazo were at the port to meet the ship and supervise the offload of the sub into a secure storage facility.

As relieved as Vescovo was to have the crown jewel of the expedition safe, he was becoming increasingly concerned with the mistakes and missteps that had piled up before the first deep dive had been attempted. The delays had also caused the dedicated science week that Vescovo had promised Jamieson to be scrapped. Any science done in the Puerto Rico Trench would have to be lumped in with the ocean floor mapping and the dive days.

"We have a long way to go to execute this mission, and so far, a lot of things haven't gone right," he said. "We can rack them up to weather, human error, bad luck, but I am getting very concerned because too many things seem to not be going our way, too often."

He, like everyone involved, hoped to soon turn the corner, starting with a successful dive to the deepest point in the Atlantic Ocean.

CHAPTER 11

OFF THE DEEP END

The Five Deeps expedition felt like it was at the crossroads. The *Limiting Factor* was heading into its first deep ocean dive, the Puerto Rico Trench in the Atlantic Ocean, but no one was sure if they would be able to achieve the goal. They hoped they could. They thought they could. Certainly, after three years of planning, two sea trials, and an open water dive in the North Atlantic, they should have been sure. But based on their rocky track record and series of engineering setbacks, they weren't sure they actually could.

On December 12, the night before the *Pressure Drop* ported in San Juan, the Triton team gathered for dinner at an Italian restaurant in their hotel and discussed whether the sub was ready. Most agreed that it was, but there was some hesitation, as it had not been back in the water after spending six weeks in the shop, during which time it had been disassembled.

"Man, I fuckin' hope we get this right," Lahey said, sipping red wine. "No, we *will* get this right."

The *Pressure Drop* came alongside at Puerto Nuevo Terminal in the industrial area of San Juan Harbor the following morning at 6:00 A.M. The ground crew went to work loading the *Limiting Factor* onto the ship so that it could sail that afternoon. For all the anxiety over having to ship the sub to Puerto Rico to rejoin the *Pressure Drop* after it went to dry dock, the process went smoothly.

However, on the run from Curaçao, Captain Buckle and his first engineer had noticed that the ship's starboard propeller was operating at only 65 percent capacity. Yet, despite the compromised prop and the installation of the bulky gondola and sonar, Buckle calculated that the ship's average speed had actually *increased* since it left dry dock. One factor was the cleaning of the barnacles from the hull. The other was that the gondola had lowered the hull slightly in the water and also unexpectedly produced a sort of Venturi effect which had reduced the overall drag of the ship.

A diver was hired to check the prop while the ship was in port in San Juan, but there was no visual evidence of what was causing the prop to give off the unusual resonance which had reduced its operating speed. The repair would have to be done later, at some point before the ship set out to cross the Southern Ocean in early January. For now, the focus was the first deep ocean dive.

The *Pressure Drop* sailed at 6:00 P.M. on the 13th. Everyone gathered in the sky bar to take in the view of El Morro Castle as the ship departed the harbor. Absent was Geffen, the series executive producer, who was busy with other obligations.

At an all-hands meeting, McCallum reviewed the dive program. The schedule allowed for four days of diving. These were to include a test dive with Vescovo and Lahey, a certification dive with Lahey and DNV-GL's Struwe, Vescovo's solo dive to the bottom, and a science dive. Jamieson wasn't happy with the setup, as he had originally been promised a week of dives, but there was little he could do about it.

Lahey presented Vescovo with a desktop model of the *Limiting Factor* for his office. Vescovo thanked him. He held up the model and cracked a smile.

"It's a little damaged," he said, pointing to a nick in the side.

"So is the real *LF*," Lahey retorted, prompting chuckles all around the room.

⁂

The first thirty-six hours were taken up mapping the Puerto Rico Trench with data collected by the newly installed Kongsberg EM 124 multibeam echo sounder to identify the deepest point. Created by the shifting of the North American and Caribbean tectonic plates, the subduction zone is situated on the boundary of the Caribbean Sea and the Atlantic Ocean, roughly 500 miles off the coast. It was discovered in 1876 by Great Britain's HMS *Challenger*. The trench's deepest point had been identified, but not verified by physical visitation, in 1939 by an echo sounder on the USS *Milwaukee*, hence it had been named the Milwaukee Deep. The most recent manned dive to the deeper depths of the trench had been done in 1964 by the French submersible *Archimède*, but it was a science mission and was not focused on verifying the deepest point. Based on the early data and more recently published surveys done using remote operating vehicles, both Jamieson and geologist Heather Stewart posited that there was, in fact, a deeper and unvisited point to the east of the Milwaukee Deep.

Cassie Bongiovanni, a hydrographer, had been hired as the expedition's lead mapper. For the native Texan, it was something of a dream job, as she had recently completed her master's degree in Ocean Mapping at the University of New Hampshire's School of Marine Science and Ocean Engineering. Though only in her mid-twenties, she had worked with NOAA's mapping efforts in the agency's Integrated Ocean and Coastal Mapping (IOCM) department after Hurricane Sandy, worked with autonomous mapping vehicles in the Antarctic, and helped map seeps off the west coast of the U.S. Bongiovanni would be on board to produce the first detailed 3D maps of all the trenches visited by the expedition.

Stewart would review the new maps and compare them to existing images. In addition to finding the deepest point, the goal was to collect more data on the area in the form of high-resolution images for further study of the trench. Ideally, the sub would also be able to collect rock samples using the manipulator arm and place them in the landers' baskets.

A massive earthquake in the trench in 1918 had caused a destructive tsunami and chances were that another would happen again in our lifetimes. "We just don't have much data on the trench, so predicting when and if there will be another tsunami is guesswork," Stewart said. "We're hoping to begin to advance that study with the mapping, image capturing, and samples."

⁂

The first dive was set for December 15 at 8:30 A.M. As the Triton team ran through its prelaunch checklist of the *Limiting Factor*, the Atlantic team readied its cameras. Four were mounted inside the sub. One cameraman would be on the ship, one in the *Xeno*, and one operating a drone. The four new underwater cameras capable of capturing high-definition images, two on the sub and on one each lander, had come from Deepsea Power & Light, a Triton vendor.

"Our challenge is that on the underwater cameras, none of the footage just plays easily, and none of the footage works in our editing system straightaway," Atlantic's Ian Syder explained. "So to convert it all and make it viewable, we have built an edit system on board. The footage also requires real-time conversion."

Despite the fact that the Triton team had been working on the sub for more than a month and a half, six hours after the scheduled time the sub was still not ready to launch. Lahey was at wits' end as small problems reared themselves, ranging from electronics that wouldn't start up to the levers operating the A-frame crane needing WD-40.

"That's what fucking happens with these subs," Lahey said, as he barreled across the deck to troubleshoot yet another niggling issue.

Finally, at 3:00 P.M., Lahey and Vescovo climbed into the *Limiting Factor* for a dive to 1,000 meters to test its systems. The sub didn't make it past 100 meters. On its descent, water began leaking in, noticeably, through the finely machined hatch. The situation wasn't life-threatening, but it wasn't diveable either.

The science team, which had risen at 6:00 A.M., was more successful with its lander deployments. The landers recovered tubes full of arthropods and a hagfish, a slimy sea creature that has no skull and breathes through its anus. A camera on one of the landers captured a two-foot-long shark with beady eyes attempting to dine on the hagfish, only to have the hagfish eject slime into the shark's mouth and scare it away. People took turns gathering around Jamieson's computer and watching.

That evening, as the sunset over the nearby uninhabited Desecheo Island produced a sweeping rainbow, the Triton team went to work fixing the leak. Magee shaved ¼ millimeter off the "O ring" housing around the hatch to get a better seal. He then boarded the sub on the deck. Lahey closed the hatch and filled the trunking with tap water. There was no leakage. Lahey was convinced they had fixed the issue and that the sub was ready to go.

But the following day, the 16th, several issues again delayed the launch, creating frustration all around. A panel covering cables running from the outside of the sub through penetrator plates to the inside was missing, apparently lost at sea during the previous day's abbreviated dive. Magee tied the wires together to secure them. "Panel's just cosmetic," he said.

On the electronics front, Blades had his hands full. The forward camera on the sub was not working, and there was no GPS signal. Two of the landers were also out of service because of modem issues.

When Vescovo indicated to Lahey that Blades needed to fix the landers since they were needed for navigation on the bottom, Lahey curtly said, "Tom is focused on the comms for the dive, and we'll address the landers afterward." Blades chimed in, "I'm absolutely sure I can fix it, just not in the next half hour."

The dive with Lahey piloting and Vescovo riding shotgun finally got underway at 3:15 P.M. It was aborted five minutes later because the hatch was leaking slightly—yet again.

Though the launch had gone fairly well, the recovery was messy. Because the *Limiting Factor* surfaced so quickly, the ship had to reposition itself. While this was occurring, the sub was left pitching from side to side in the choppy water for over thirty minutes. Then, after the swimmer, Colin Quigley, had attached the main line and the two side tag lines to the ship's crane, one of the tag lines snapped. Without both tag lines, the deck crew would have no way to fully stabilize the sub when it was raised out of the water by the A-frame crane.

Frank Lombardo, one of Triton's deck hands and maintenance techs who has a "failure will not happen on my watch" attitude, radioed for the *Learned Response* to pick him up. The tender zipped back to the ship. Wearing no life jacket, he jumped into the support boat and rushed out to the sub grasping a new tagline. Without hesitation, the Floridian and expert diver dove in the water and clipped the tagline to the sub. He then swam back to the support boat.

The recovery process took more than an hour, during which time Vescovo—for the first time in his life caused by rough seas—lost his lunch.

When the sub was finally in position behind the ship, Vescovo climbed out of the hatch and grabbed the side railing attached to it for entrances and exits. Just as he steadied himself to step into the tender, a white-capped wave struck the sub, and he was tossed into the water. Quigley, who had already broken a rib the day before, dived in after him, as Vescovo swam to the tender and climbed in.

Minutes later, Vescovo was back on board the *Pressure Drop*. His fireproof pilot suit was soaking wet. Typically, after each dive he would give a brief on-camera interview to the Atlantic film crew, but this time he brushed by them without saying anything. He walked through the control room with barely a pause and down the hallway. As he made a turn into the stairway to go to his cabin, he left behind wet footprints and a trail of disappointment with everyone involved.

As he entered his cabin, one of the stewards was restocking his supply of Diet Coke. The steward took one look at the dripping wet Vescovo and asked, "Are you okay?"

Vescovo managed a smile. "I've had better days."

Meanwhile, with Lahey now back on deck, the Triton crew was struggling to maneuver the sub in the right position to lift it out of the water. It was bobbing around just inches from the stern as Magee tried to tighten the tag lines and stabilize it before hoisting it onto the ship. After a few harrowing moments, they succeeded.

Once the sub was on deck and locked in place, Lahey yelled, "Nice fuckin' job Magee!" Lahey, Magee, and the two crane operators hugged. As poorly as the entire day had gone, disaster had been averted and the *Limiting Factor* was safely back in its cradle.

The atmosphere at dinner was surprisingly light. Lahey sat with Vescovo, and the two discussed the leaky hatch and a problem with the main ballast tanks that needed to be corrected. Lahey, who was still inside the sub when Vescovo fell into the ocean, joked that when he saw

Vescovo soaking wet, he thought that he had gone for a swim because it was such a beautiful day. "Well, the water was warm," Vescovo said.

As he left the table, Lahey turned serious and threw some of his typical optimism on the problems. "We're gonna get it right, do it again tomorrow and get the trench done."

"That's all I want for Christmas," Vescovo said.

⁂

On Monday, December 17th, Lahey and Vescovo boarded the sub for a third test dive. This one really needed to be successful, as there was only one more diving day left on the schedule for Vescovo to attempt his solo dive.

The *Pressure Drop* had moved to the leeward side of the island, where the seas were calmer. Though the launch was uneventful, the hatch began to leak as the sub descended. This time it was more like seepage than a trickle of water, so Lahey and Vescovo decided to continue the dive.

When they reached the bottom, at 1,000 meters (3,280 feet), they radioed that they were going to look for the lander. Kelvin Magee, who was serving as the comms officer, smiled. "Small victories," he said to the anxious room. "We need some."

Inside the sub, Vescovo was piloting and things were going well. The water ingress from the hatch had stopped as the water pressure had evidently sealed the hatch. He and Lahey were marveling at the exotic sea creatures they could see through the viewports. This was what a dive was supposed to be like.

"Should I try the manipulator arm?" Vescovo asked.

"That's up to you," Lahey replied.

"I just want to see if we can get it out," Vescovo said.

"We can definitely get it out," Lahey said.

Vescovo powered up the hydraulic system, the manipulator controls, and then grasped the arm's control lever and extended the arm. Lahey was pressed against the viewport, watching intently. As the massive hydraulic arm with a claw on the end came into view and extended away from the sub, Lahey said, "There's your arm." As Lahey turned away and prepared to make a call to the surface, Vescovo maneuvered the arm for a few seconds and began to re-cradle it for ascent.

Suddenly, there was a snapping noise, followed by a small puff of sea dust coming off the seafloor.

"What the hell was that?" Lahey said.

"We just lost the arm," Vescovo said.

Lahey's eyes widened. "No . . ."

Without saying a word, Vescovo pointed for him to look outside his portal and see the detached arm. "Oh my God . . ." Lahey lamented.

Vescovo stowed the arm's control lever and leaned back in his seat, utterly deflated. "I don't know where we go from here, Patrick."

Without the significant weight of the manipulator arm, the sub began an unplanned ascent.

"Surface, we just lost the manipulator arm," Vescovo radioed. "We are on our way up."

As Vescovo's voice filled the control room, some people put their heads in their hands, others' eyes widened in disbelief. One person choked on his soda. No one said anything.

Magee, who was not expecting a comms check for another few minutes, picked up the handheld microphone. "Roger that," he said in an even tone. "Left bottom without the arm."

Struwe knew the cause right away. The frangibolt holding the 100 kilogram (220 pound) arm had snapped. During the design phase, Struwe had warned against using a single bolt for the arm because it might not be able to support the side to side movement once the arm was extended. Though it wasn't an "I told you so" moment, it was an expensive one: a new arm would cost in the neighborhood of $300,000, and months were needed to make it.

The Triton team went to the deck to recover the sub as the film crew rushed to position their cameras and capture the emotions of Lahey and Vescovo. The sea state had picked up since the launch from a 1 to a 2-plus. It took the LARS team twenty minutes to position the sub behind the ship. Vescovo and Lahey disembarked and returned to the ship in the *Xeno*.

Once in place, the sub was hoisted by the A-frame crane, but it wasn't high enough out of the water when the crane came to a stop. At that precise, inopportune moment, a tailwind whipped through and rocked the sub into the stern of the ship, crushing two thrusters. As the sub bounced off of the ship, Magee, who was running the deck operations, shouted: "Up on the main! Up on main!"

The sub was quickly pulled above the ship, locked into place and moved onto the ship.

Lahey went down to examine the damage with Magee. The two mangled thrusters would need to be replaced. This in and of itself wasn't a major undertaking, as good planning had assured that there were spares available on the ship. The bigger issue was that a post-dive inspection revealed that a handful of the electrical penetrators into the sub's capsule needed to be replaced, and that was not easily done on the ship. It was a significant issue. In the current condition, the sub couldn't dive.

This new and difficult repair, plus losing the arm, was a totally demoralizing set of events. "It's absolutely heartbreaking," Magee said. "It could be curtains for this trip."

Buckle was becoming increasingly frustrated with Triton's inability to execute the launch and recovery system, and was again bemoaning that there was not a man-rated crane, as he had done when he first saw the ship.

"Patrick always thinks he's right," Buckle said after the disastrous recovery. "You can't tell him what to do. Now we have a launch and recovery system that doesn't work. The fundamentals are flawed. Either the sub is too big for the A-frame, or the A-frame is too small for sub. You can mitigate the problems but you can't remove the problems, but if we have a new LARS, we can."

The ultimate solution was to install a man-rated crane. The problems with that were cost and time. The cost would be in the $1 million range, and the time to install it would suspend the expedition, almost certainly causing it to miss the Southern Ocean dive window. Triton would simply have to figure out modifications and process changes to make what they had work. This is what Vescovo did in his day job with factories, and this was a similar problem.

However, Vescovo was also frustrated with the continued series of mechanical issues with the sub, particularly with the problems with the all-important VBT system. "I kept telling them I wanted an M-16 not an AK-47," he said that night at dinner, employing a rifle metaphor. "The M-16 is a better rifle, a precision firearm that is technologically more advanced. The AK-47 is not as accurate, though it also gets the job done."

But while he was slowly losing his patience, he still maintained his sense of humor. "I've joked that this sub cost me an arm and a leg," he added. "Today, it literally cost me an arm." He quickly added that he had no intention of paying for a new one—that since its loss seemed to be a fundamental design flaw, it was Triton's problem to warranty. If they couldn't, he would just do without it and focus on diving, and not potential sample collection.

Lahey had reached a breaking point of sorts. After dinner, he met with his team in the control room. Morale, which had seen its valleys, had crashed like an ill-equipped website. "There was a collective sigh that we had reached the end, that this was not going to happen," he later recalled.

That night, Vescovo, McCallum, Buckle, and Lahey met to discuss what to do next. The schedule could be extended, messing up travel plans heading into Christmas, but that wasn't really the issue. The fact was, there was little evidence that the deep ocean dive could be accomplished with all the problems that had occurred. More than that, it felt like the entire expedition was in jeopardy because the sub couldn't execute a successful dive.

"Have I just wasted $25 million?" Vescovo asked Lahey at one point. "I'm prepared to walk out of here and write it off as a bad debt if there are major design issues that can't be fixed."

"You can't do that," Lahey pleaded. "We can fix it. You've got to give me more time. I need one more day. We can do it."

The group began openly debating the short-term alternatives, either packing it in or giving Triton a day to fix the sub and attempting the deep dive. McCallum gently pushed back, talking about how far the team had come since sea trials. Lahey pleaded for the next day to try and make the sub diveable. He believed in that period of time they could replace the thrusters, reroute electrical connectors through those used by the now-absent manipulator arm, and fix the other issues. Buckle figured there was no downside to at least letting Triton see if they could make the sub ready.

After everyone weighed in, Vescovo said he wanted to think about it.

McCallum then met with Vescovo alone and gave things a hopeful spin. "You can't pull out now," he said. "This has never been done before. We are operating out here without a playbook, without a recipe, without a template. We are learning as we are doing. We are breaking as we are doing. My list of issues from sea trials 'til now has gone from over 100 to 12. But you can't keep leaning on Patrick like that because you are going to give him a stroke and that's not going to help us."

Vescovo, sometimes reasonable to a fault, made the logical decision. Just as he had decided that the team needed to take a risk at the dive off the *Titanic* site to see if the sub could be launched and recovered in rough seas, he had to try for this dive. He remained confident that the absolute *core* systems of the sub—its life support system, pressure hull, and ability to go up and down the water column—had always worked well and had not been compromised by all the failures in the past few days.

"It's a calculated risk, but I've taken greater risks," he said. "This one is reasonable."

He would give the Triton engineers the following thirty-six hours to make the sub diveable, and if accomplished, he would attempt his first solo dive ever on Wednesday to the deepest point of the Atlantic Ocean. Either way, the Puerto Rico Trench mission would end on Thursday, and if that happened without a successful dive, who knew what would happen next with the expedition.

~

On the morning of the 19th, the sun was radiating off the Atlantic Ocean in a way that made it look like light reflecting off a massive diamond. Overhead, the sky was azure with a slight dusting of clouds visible on the horizon. The sea rippled slightly but was the calmest it had been the entire trip. The conditions to dive were ideal.

After Triton had worked virtually nonstop for thirty-six hours to fix the seeping hatch, replace the damaged thrusters, and repair the VBT system, Lahey deemed the sub ready to go. The big fix was that the conductors freed up by the loss of the manipulator arm allowed Blades to restore full functionality to the other electrical systems—though to make the connections work, there was now a cable running across the back of the two seats.

"We completely rewired the sub," Lahey said. "I don't want to use the term 'hotwired,' but it's ready."

All of the teams gathered for a morning meeting. McCallum started by delivering a pep talk. "We have all worked very long and very hard to get to this point," he said. "It's been a long road . . . the culmination of a dream that's become a reality. In a normal speech, I would call this the moment of truth. But I don't think it's that. I think this is the moment we have been working toward. I would call this the moment of harmony, when all elements come together."

He outlined the dive and detailed the assignments. "We just have to go and execute," he said. "The plan is pretty simple . . ." He paused, gave a heavy laugh, and added, "He says preparing to send a submersible to 8,400 meters under the sea."

He then showed everyone the Explorers Club flag that had been presented to Vescovo in New York. "We are going to add a lot more history to this flag." He folded the flag and handed it to Vescovo. "Take the flag down and bring it back."

"It's like mountaineering," Vescovo said. "It only counts if you come back."

With that, the meeting broke up, and everyone went to prepare for a dive that had been more than three years and tens of millions of dollars in the making.

Vescovo returned to his cabin and packed a bag of items he wanted to take to the bottom of the ocean. For someone who is not sentimental by nature, these were objects that reflected his life. They included the jacket he took to Mount Everest, the North Pole, and the South Pole. He had several folded flags: an American flag given to him when he retired from the Navy reserves, flags of his alma maters—Harvard, Stanford, MIT, and his high school—as well as a UN flag, the Explorers Club flag, and an Albanian one, a tribute to the homeland of his longtime girlfriend, Monika. He also had a Cartier "Love Bracelet" for her, the first of five that he would take down to the bottom of each ocean and then have the coordinates and depth engraved on each one. He had bracelets for his sister Victoria and his step-nieces and two watches, his father's gold Rolex and a silver Rolex his deceased father had given him when he was eighteen. On his actual wrist, however, he wore his new titanium Omega Seamaster Chronometer. He also had the dog collars from his three Schipperkes who had died over the years.

Vescovo zipped his "Go bag" and headed for the launch deck. With Lahey, he walked around the sub and ran through the pre-dive checklist. Finding nothing of concern, he proceeded to the tender to board the *Limiting Factor.*

"Safe dive . . ." Lahey said.

"I'll try," came the reply.

Like clockwork and with precision, the sub was launched; Vescovo boarded, and the swimmer Quigley sealed the hatch and released the main tow line. All systems in the sub were operational and the control board was green. The dive was a go.

"Life support good," Vescovo reported from inside the *Limiting Factor.* "Starting pumps . . . see you on the other side."

With that, Vescovo began to descend without a copilot for the first time. Lahey had never been crazy about him diving alone, especially not on his first attempt to the bottom. "I'm as nervous as a whore in church," Lahey said.

As the dive progressed uneventfully, everyone in the control room seemed curiously optimistic. Things were going just as they had been drawn up. Vescovo performed his comms checks on the quarter-hour, giving his depth, heading, and indicating that life support was good.

After twenty minutes, he passed through 3,000 meters. An hour into the dive, as he crossed 5,000 meters, he noticed a subtle chirping sound in the capsule. "What the hell is that?" he said to himself. He began cycling circuits and eventually the strange noise stopped.

During this process, he missed a quarter-hour comms check, or at least the control room couldn't hear him. Lahey radioed the ship's bridge to contact the *Learned Response* to come to a stop to prevent any interference.

"Victor . . . can . . . you . . . read . . . me . . ." Lahey said, pausing for a reply. Hearing nothing but static, he hunched over and uttered, "Oh my god . . ."

Lahey reached for a headset and pressed his hands over the ear pieces. Seated next to him, Blades began adjusting knobs on the control board and assured Lahey everything was okay. Almost twenty-five minutes had passed since they had heard from Vescovo.

"Victor, can you read me?" he said slowly. "I say again, can you read me . . . over."

After a full minute of silence, Vescovo's voice came across. "Roger . . . have you good," he said. "I . . . will . . . talk . . . louder."

Lahey ripped off the headset. "Fuckin' A . . ."

McCallum put a hand on his shoulder. "Breathe . . . breathe . . ."

As Vescovo continued his descent, the comms became clearer and depth records began piling up. Reaching 5,000 meters (16,404 feet) made him the second solo pilot to go that deep—only James Cameron in his *Deepsea Challenger* had made a deeper solo dive, to full ocean depth, recorded at 10,908 meters, in 2012. Passing through 6,100 meters (20,013 feet) made the *Limiting Factor* the first American-made submersible to dive deeper than that since the U.S. Navy's DSV-4 *Sea Cliff* in 1985. At 7,100 meters (23,200 feet), it surpassed the world's deepest diving active submersible, China's *Jiaolong*, which had reached 7,020 meters in 2012.

When the *Limiting Factor*'s sonar altimeter indicated the Vescovo was near the bottom, he began looking out the viewport in anticipation. With 200 meters to go to the seafloor, he began to release his small VBT weights to slow the descent and become neutrally buoyant. As designed, the sub slowed as it approached the bottom. At 30 meters, he noticed that the down viewport was actually bringing in more light—the powerful lights of the sub were subtly reflecting off the bottom. Quickly, as the sub closed the last 10 meters, Vescovo could make out what looked like a tan moonscape below him. And then, with barely a sound and imperceptible downward motion, a small puff of silt rose around the *Limiting Factor*.

He had touched bottom.

"Surface, *LF* . . . present depth 8,589 . . . at bottom . . . repeat . . . at bottom" he said, reporting the depth being calculated in decibars.

The control room erupted in cheers. Lahey threw his arms up in the air and made the rounds, high-fiving each one of his team. When converted from decibars to meters, Vescovo had made it down to 8,376 meters (27,480 feet), or 5.2 miles, the bottom of the Puerto Rico Trench, achieving his goal of becoming the first person to reach the absolute nadir of the Atlantic Ocean at 2:55 P.M. EST on December 19, 2018.

Vescovo piloted the *Limiting Factor* around the bottom to see what could be seen and film the surroundings with the high-definition cameras affixed to the sub. He was struck by the fact there was virtually no current and the brownish sediment on the bottom was barely shifting, creating an elegant geometric pattern of ripples on the seafloor. "I'm at the bottom of the Atlantic Ocean," he said to himself. "It's like I'm on another planet."

With one hand on the joystick to engage the thrusters, he peered out the viewport. "Looks like the surface of the Moon . . ."

Then he spotted an object dead ahead. He piloted in that direction to see what it was. Blackish in color and round in contour, it appeared to be a cracked oil drum wedged into the sediment. He frowned and sighed disgustedly. "Really?" he muttered.

After Vescovo had been roaming around the bottom for forty-five minutes, Lahey began to grow nervous. "We have to stop for safety reasons," he told Blades who was sitting next to him running the electronics—though there was no obvious safety issue.

He radioed Vescovo, recommending that the dive end. Following protocol, Vescovo replied that he agreed and was dropping the surface weight. There was a pause, a sharp "click" in the capsule as Vescovo flipped the weight release switch, and then he was, indeed, ascending. The sub was on its way up.

"Roger that, surface weight dropped, positive ascent," Lahey said into the handheld microphone. "Congratulations."

Lahey clicked off and spun around in his chair. "How about that? You fuckin' guys rock. . . . Now let's get him back on board."

Two and a half hours later, at 6:17 P.M., a burst of light breached the water's surface on the starboard side of the ship, followed by the white top of *Limiting Factor* surfacing to a backdrop of grayish clouds with sunlight streaming through them and lighting the water a surreal shade of aqua.

The LARS team swung into action. Fifteen minutes later, after a flawless and textbook recovery, Vescovo popped out of the sub. "One down!" he yelled, holding up his index finger.

"Victor Vescovo . . . rock star!" shouted Lahey, pumping his fist.

Everyone on the deck applauded. Forty-eight hours after the Five Deeps Expedition appeared to be unraveling, it had achieved its finest moment to date.

~

That night, a celebratory barbecue was held in the sky bar. Vescovo clutched a bottle of Veuve Cliquot, drinking directly from the bottle, and detailed his experience of being the first person to ever see the absolute bottom of the Atlantic.

The great irony was that losing the manipulator arm had made the dive successful. The conductors freed up by not having an arm ended up providing Blades a method to fix all of the other electrical faults by rewiring the sub.

"My wife always says, 'Not all bad things come to hurt you,'" Lahey said. "There is something profound about that. We were sitting there facing what looked like an absolute failure when the arm dropped off, but that was what made it possible for Victor to have that dive."

Lahey also pointed out that pushing for the extra time to fix the sub when it looked like all was lost had saved the entire trip. "People criticize my optimism, but Victor was ready to throw his toys out of the crib," Lahey said. "It was my fucking relentless optimism that pushed us through and got that dive done."

There was also talk of what had been discovered by the science team, though the footage from Atlantic's cameras had not been fully reviewed. Science had taken a back seat to the core mission—to dive to the bottom—but at least four new species of sea creatures had been discovered, and they could be used as a benchmark in the ongoing examination of biodiversity at full ocean depth. On the geological side, Stewart had identified "slumps" in the trench on the sonar scans, small depressions that could trigger a tsunami as they expanded, data that she would turn over to local geologists in hopes of furthering their understanding of their undersea environment. "Baby steps, but a start," Stewart said.

As the champagne flowed, an emotional and spent Lahey stood to present Vescovo with a submersible test pilot certificate. "I've trained a lot of sub pilots in my life, been doing this thirty years," he said. "Never have I had a trainee who has had to endure so many problems over the course of a dozen dives. It has been truly surprising, unnerving, disappointing, but at the same time pretty remarkable. It has been an opportunity to test you in ways that maybe you hadn't expected."

Vescovo took the certificate and hugged Lahey.

"Today, you really stole the show," Lahey continued. "I think everybody on the ship thought, Holy shit man, this guy has balls of fucking steel . . ."

Over the laughter, someone shouted, "Titanium!"

Vescovo stood and raised his champagne bottle. "Thank you, Triton," he said. "Thank you . . . you guys made it happen."

There was still a long, long way to go to complete the next four ocean dives, but in one single, record-breaking dive, it felt like the entire expedition had turned a corner, that possibility was heading toward reality.

CHAPTER 12
MEDIA MATTERS

The morning following the *Limiting Factor*'s record-breaking dive to the bottom of the Atlantic Ocean and the champagne-soaked celebration that extended into the wee hours, on December 21, the *Pressure Drop* anchored in the harbor near Samaná, a small resort village in the Dominican Republic. The tender transported Vescovo ashore so that he could fly his plane to Miami in time for a midday board meeting conference call, and then be back in Dallas for more meetings that week.

Vescovo landed in Miami just after 1:00 P.M. He was greeted by a forceful, threatening email from Mimi Gilligan, the Director of Commercial and Legal Affairs for Atlantic Productions, as well as several from the expedition team, all of whom had received the same email.

Gilligan wrote that "a number of organizations which are part of Caladan's Five Deeps Expedition have violated the contractual agreement that Caladan has signed with Atlantic Productions which grants Atlantic Productions all media rights, including photography, in the Expedition."

In the email, she requested that all parties, including EYOS and Triton, clear any photos with the Richards Group, the marketing firm hired by Vescovo, before disseminating. "It also needs to be made clear that any person associated with the crew, whether they have signed or not signed the photography agreement, will cause a violation of Caladan's media contract with Atlantic if they share any Expedition photos or footage with the public (including on personal social media)," she wrote.

In conclusion, Gilligan threatened litigation. "We will give all parties 24 hrs to remove all non-approved material from any websites, social or other media," she wrote. "Afterwards, both Atlantic Productions and Discovery will seek appropriate legal recourse, using whatever course of action is most appropriate, to address any violations of the media agreement."

The letter had come as a result of Triton posting news on social media of the dive and McCallum reposting the news trumpeting the record-breaking dive that everyone had worked so hard to accomplish. But after the high of the dive day, the letter from the film company threatening legal action against several people on the expedition hit like a cyclone.

While in transit, McCallum emailed Vescovo that Atlantic was being heavy-handed and undiplomatic. "Welcome back to Earth," he wrote. "I'm not sure how she expects us to honor an agreement we've not seen (let alone signed)." He also complained that there had been no immediate publicity about the dive from Atlantic or Discovery. He then emailed Gilligan that he had no agreement regarding media with Atlantic, Discovery, or Caladan. "Your threat, while hollow, is irritating as it is counter to the culture we have been steadily building over the last year," he wrote.

Geffen defended Gilligan's terse email by saying that it was necessary to protect the media rights to the expedition going forward and ensure the value of the expedition to all media partners and sponsors. "Since Triton and others were not abiding by the contractual agreement Victor had made to us (and us to Discovery) and this had been going on for some time, there were limited options to doing this differently," Geffen wrote. "It was a time-sensitive issue and thus couldn't wait until celebrations were over and people were back at their desks. And it wasn't a new requirement that should've caught anyone unaware."

Vescovo agreed with McCallum that Atlantic was being overly aggressive and was upset that the first reaction by Atlantic was not to call him and work out a resolution, but to threaten immediate legal action against the key members of the team—including himself. However, he was also dismayed that EYOS and Triton had apparently and completely ignored his earlier instructions to them that all media releases first be approved by the firm he had hired to coordinate just this activity. Nevertheless, he emailed all the team leaders that he would speak to Geffen and Gilligan about "the uncalled for, immediate threat of litigation." But he made it clear that as the financier of the expedition, he was seeking cooperation from all subcontractors on the expedition to work with him and Atlantic to use only approved materials and to allow the Richards Group to act as the gatekeeper of all information being released about the expedition. He asked everyone to be patient with the release of information and pointed out that there was little benefit—and possibly even damage to the expedition—by everyone pursuing their own, independent media desires. Vescovo was also mystified that there was this mad rush to publish media content when a few days to polish any press releases and coordinate actions made no difference to the public at large.

"I am not trying to directly control others' media use, but am asking that since I am paying for everything, that my desire to only use approved material is respected," he wrote to all parties. "That all said, I will do my best to force better understanding from Atlantic, and also try to get everyone to continue to use the Richards Group as the source for approved material dissemination. If there are absolute refusals from anyone to do this—to work from a common PR repository that we all try to manage—then that will pose some challenges and decisions for me."

What seemed like a prickly, yet somewhat minor issue that could be easily resolved would soon pick up steam and threaten to break up the expedition team.

❧

On December 21, the Richards Group issued a press release cleared by Atlantic on the record-breaking dive. The news received a smattering of online coverage from several outlets, including the "For the Win" section of *USA Today*, businessinsider.com, the *Guardian*'s (UK) website, lifescience.com, and a few others. A detailed story appeared the same day on *Popular Science*'s website, written by Josh Dean, who had been on the ship for the dive. But by no means was the dive big news.

Vescovo, who wasn't doing the dives for press attention, was not surprised by the tepid response and was fine with the coverage. "Look, I get it," he said. "People hear about this and say, 'That's neat, but how does it really affect me?' We are looking to open some doors of exploration, and we have opened one."

What also became clear as the saga over the restriction of the release of information began to involve lawyers in the ensuing days and weeks was that the expedition team was at its best when it was on the ship sharing a myopic vision to achieve a common goal. The *Pressure Drop* was the team's bubble, a floating village where men and women at the top of their professions had come together for something they believed in and cared about. It was a haven where little else existed beyond achieving the goals of the mission.

On each of the four trips at sea to date, most of the conversations on the ship centered on the object of everyone's affection, frustration, and possibility, the *Limiting Factor*, and what the sub had done, what it needed to do, and how to work in harmony to make its dives successful. Even the problems plaguing the sub during the Puerto Rico dives that threatened to upend the expedition—and Vescovo's somewhat impetuous threat to pull the plug altogether—were worked through by everyone coming together and addressing the problems at hand.

Off the ship, however, every logistical complication and internecine squabble seemed to magnify. Certainly, there had been many—the *Limiting Factor* being unprepared for sea trials, the questionable scheduling of the failed *Titanic* dives, the sub being forced to undergo a near-total refit, the underwhelming media launch day, the turmoil over the installation of the sonar, and now the threat of legal action from Atlantic Productions. But when everyone stepped back and reflected on how far things had come and what the journey meant thus far, the tone of the discourse turned positive and productive.

This was evident two days after the press release was issued. After returning home and resting up, Lahey sat down and typed up a lengthy email to Vescovo on December 23 to wish him a Merry Christmas and to articulate what the Atlantic Ocean dive had meant to him and his team. In the email, he didn't even mention the debate over media issues that was already in the hands of the lawyers.

"This most recent deployment was among the most difficult of my life," Lahey wrote. "We had a seemingly endless number of issues, but the team held fast and we made it happen in the end. I have never been more proud of a group of people than I was at the conclusion of this series of dives. . . . Most of all, I have such tremendous admiration for you and your willingness to make this historic dive to the deepest point in the Atlantic Ocean as your first solo. . . . It was my pleasure and privilege to be present as you made history."

Vescovo replied: "If it wasn't obvious, I quite literally trusted you and your team with my life. One can imagine the trepidation some might have in getting into a prototype submarine that had a leak in the hatch on its previous dive, and then taking it down, solo, to 8,400 m. But I hope it shows my faith in Triton and you personally that I would do so. We don't always completely agree and we have shared some frustrations, but I always have enormous respect for you and your team's capabilities."

The stronger the bond over the shared goals could become, the better the chances the expedition had at succeeding. But first, it had to hold together.

~

As the debate simmered over the media problem, there was a nagging issue with the permit for the Java Trench dive in the Indian Ocean, scheduled for late March 2019, that needed to be addressed. Because the area was within the Exclusive Economic Zone (EEZ) of the Indonesian government, the expedition needed a permit to conduct scientific research and to collect any marine life. McCallum had hired Andy Shorten, a Bali-based expert in maritime matters to help facilitate the permit, but things were not moving forward. Vescovo decided to fly to Jakarta over the holidays and meet with the Indonesian authorities personally to try and get things moving.

The issue had first reared its head in July when McCallum applied for the permit. Vescovo's name was on the application as owner of the *Pressure Drop* and the *Limiting Factor.* After receiving the request, the Indonesian authorities apparently conducted a background check on

Vescovo and discovered that he had been a reserve U.S. Navy intelligence officer. A combination of this and his mysterious full ocean depth sub threw up a red flag.

McCallum emailed Vescovo tongue in cheek that "the Indonesians are onto you." He asked if there was anything he or Shorten needed to be aware of.

Vescovo laid out the facts. He had retired from the Navy in October of 2013. During his career, he had never served in Indonesia, nor was he ever a member of any military exercise that involved Indonesia or any of its neighbors. He was a reserve officer for his entire career, and after 2003, typically only spent three or four weeks per year on active duty service, and then only in a training or military exercise capacity. From 2008 to 2011, he was attached to the staff of the commander, U.S. Seventh (Pacific) Fleet, but had no assignments that directly related to Indonesia during that time and served as a military trainer and analyst based in the U.S. He had only visited Indonesia twice. In 2004, he went to the factory of a company that manufactured eyeglass lenses outside of Jakarta that he partially owned. Then in 2011 he had visited Bali at the time he climbed Carstensz Pyramid.

To allay any concerns, Shorten set up a meeting with Vescovo and the Indonesian officials. On December 28, Vescovo made a 48-hour turnaround trip all the way from Dallas to Jakarta to give a presentation on the expedition and personally address any concerns.

Both Indonesian military and civilian officials attended the meeting. They expressed surprise—shock almost—that Vescovo was funding the entire expedition himself, without any government assistance. They asked repeatedly if he would allow one or two of their scientists to come on board and join in the science mission, which Vescovo assured them they could.

Then they showed several lengthy PowerPoint presentations—which were not in English. Vescovo politely watched, smiled, and posed for pictures with the officials at the end. He was told by a senior civilian official that the government would need three months to process the permit, and that Vescovo would need to sign an agreement that stated anything recovered would go to an Indonesian lab, a condition he agreed to.

The people he met with were all smiles and shook his hand vigorously when he departed, leaving Vescovo with the feeling that the government would ultimately grant the permit. The fallback position, however, was to dive without a permit. Technically, if the expedition didn't retrieve anything from the ocean, then the Indonesians couldn't stop them from diving because according to the Law of the Sea, they would not be conducting "marine research"—they would simply be diving in the water and testing their equipment, which was not prohibited activity. Furthermore, the law's provisions do not allow a government to stop "normal passage or loitering" in or even under the sea. Of course, without a permit, the Indonesian government could make life difficult for the *Pressure Drop* on entry and exit, as it was scheduled to ingress and egress from Bali.

❧

The back-and-forth over who could say what in the media dragged into mid-January as everyone dug in their heels. It escalated to a point that threatened to actually replace two of the teams, EYOS as expedition leader and Triton as the operator of the LARS. The situation was made more complicated because it wasn't a black or white issue, but Vescovo had no desire to get into litigation that had been threatened by Atlantic's Gilligan.

Everyone involved in the expedition had been under a Non-Disclosure Agreement (NDA) with Caladan from the beginning, though only Caladan, meaning Vescovo, had signed the media agreement with Atlantic. None of the other parties had even seen it. The reasoning was that they were all under the NDA, and therefore could not say anything publicly about the expedition. However, both McCallum and Triton took the position that the media blackout aspect of the NDA was rendered obsolete because the expedition had been publicly announced in New York. Vescovo countered that nothing in the NDAs contained any provisions that rendered it void—therefore, EYOS and Triton were making an incorrect assumption. Things escalated even further when the social media posts after the Puerto Rico dive impelled Vescovo to ask for a further, explicit agreement that they not disseminate any information on the dives without clearance.

Vescovo felt he had no choice but to do so, or EYOS and Triton would just ignore the media restrictions and *he* would be held accountable by Atlantic and the Discovery Channel. He couldn't believe this was happening, because, as he often said, he personally couldn't care less about media—he just wanted a great video record of the expedition. He thought the whole thing was ridiculous, and that everyone else was being unreasonable and completely ungrateful given the massive resources he had personally deployed to make this opportunity happen for all of them. And he had a point.

On January 11, McCallum stonewalled in an email to Vescovo. "I am not going to get into a bunfight with you about Atlantic's quest for total media domination; we have bigger fish to fry right now and it would be an insult to our relationship," he wrote—and then proceeded to do just that.

He said that EYOS did not rely on media coverage so that was not at issue. He then pointed out that he had not seen the agreement, but would review it. "You seem determined for us to sign this to protect your obligations to AP [Atlantic Productions] and DC [Discovery Channel] and imply a threat if we do not do so. So, let me review it and get a version back to you."

Vescovo was beginning to lose his patience. If McCallum didn't care about media, then what difference would it make if he signed an agreement not to do any press unless approved by Atlantic? Frustrated by the apparent hypocrisy, he told Matt Lipton that he was prepared to terminate the contract with EYOS if McCallum did not sign the agreement.

"Victor had always said, 'I need you to keep this under wraps until we are up and running,' which we had agreed to," McCallum said. "But the agreement was that after [the media event in] New York we were allowed to go 'weapons free,' which were his words."

McCallum felt that Geffen was trying to push everyone around using Discovery as his cudgel. "Anthony was operating like a petulant spoiled brat," McCallum said. "When he didn't get his own way, he was issuing these false legal threats. I thought he was a blight on the whole program. Just at the point where everybody was playing well together, everybody had this shared vision, there was a good deal of camaraderie building up steam, Anthony was operating like a lone player and everybody had to do what he wanted or he would rattle his saber and tell daddy. This was just blunt amateurism."

On January 16, Vescovo talked to Geffen and tried to persuade him to back down from the heavy-handedness of the supplemental agreement. From Geffen's point of view, Vescovo had already agreed to let Atlantic handle all media on the expedition and not to allow information to be released without his approval. This was an extension of what Geffen had agreed to with the Discovery Channel, as the network understandably wanted to control the flow of information for marketing purposes and to protect the value of the series.

"No broadcaster would allow people working on a show to post their own social media before the episodes have run," Geffen told Vescovo.

Geffen did agree that along with each official press release from the Richards Group, a certain number of photos would be made available for EYOS and Triton to use. Vescovo thought it was a fair compromise, and about as far as Atlantic could bend given the restrictions Discovery had placed on the film company.

The one thing Vescovo could not do was side with EYOS and Triton and cut Atlantic and Discovery loose over the issue. Discovery, through Atlantic, had already spent more than half a million dollars on filming. If Vescovo were to attempt to drop Atlantic, he reasoned, then he would be on the hook for that money, plus any damages Atlantic and Discovery claimed. "I cannot drop them if that is what EYOS and Triton are trying to get me to do," he said. "They would sue me for breach and put a huge number out there for damages for lost revenue from the series."

He was becoming increasingly irritated with McCallum and Lahey for not seeing this simple fact. What made it worse to Vescovo is that he felt he was paying Triton millions, and EYOS hundreds of thousands of dollars, to work for him and they were not agreeing to something as simple as a conventional media embargo. It made no sense to him.

Vescovo asked Lipton to impress upon Triton's attorney that the company had already agreed to confidentiality under the previous NDA and in the Vehicle Construction Contract (VCC), and he reasoned any release of unapproved information would constitute a violation of those agreements.

"It is much more promising to sue Triton to abide by the confidentiality in the VCC than to suffer a lawsuit from Discovery," he told Lipton. "And oh by the way, kiss the free advertising to Triton . . . goodbye, or maybe even ever working with Atlantic or Discovery again."

Though Bruce Jones was not traveling with the expedition and had been largely pushed aside, he was still spearheading the marketing of the ship and the sub, and he felt that the agreement would restrict his efforts. But Jones decided to sign the agreement because Vescovo asked them to. "We didn't have an agreement with Anthony Geffen on doing media, but my feeling was that we at Triton didn't want to stitch Victor up," Jones said. "I would have told Geffen to go fuck himself and done whatever I wanted if it hadn't been for Victor."

While Geffen maintained that Triton would receive a great benefit from the Discovery Channel series, Jones brushed that off as window dressing. "In thirty years, I can count on two fingers the number of submarines I sold because the client saw an article or a film and came to us and said, 'I've heard about you guys, I want a sub,'" he said. "There's a difference between publicity and marketing."

Lahey didn't like the restrictions either. He felt that putting out the news of the dives in the biggest possible way would be best for both the expedition and the resale of the Hadal system. "We weren't bound by any kind of agreement with Atlantic or Discovery," Lahey said. "We had never been given an agreement or asked to sign one, but out of respect to Victor, we agreed to sign the restrictive agreement because we didn't want to put him in a bad situation."

It was also no small matter that Triton would be entitled to bonuses totaling $2.5 million if certain benchmarks were met, such as diving the Challenger Deep and completing all five deep ocean dives. The best chance for that to happen was for Triton to continue running the LARS and supervising sub operations on board the ship. When they agreed, Vescovo was relieved and thankful, but it made him even more frustrated with McCallum and EYOS, because they had even less at stake than Triton.

McCallum refused to capitulate and pushed his goodwill with Vescovo to the brink. Even before this dustup, McCallum's demerits had started to add up. As smooth a hand as he provided on the ship and for all his experience, he had made what Vescovo deemed were some costly errors. McCallum had been a proponent of attempting the *Titanic* dive, despite the dates being at the edge of the weather window. He had recommended the Damen shipyard in Curaçao, which had resulted in delays, cost overruns, and possible litigation if they did not back down from some of the excessive charges.

Vescovo also felt that McCallum had not been completely transparent when he asked to bring an additional EYOS employee on board the ship for the Puerto Rico Trench dive. Vescovo agreed to let him on board because he later related that McCallum had described the person as a kind of EYOS trainee. In actuality, the person was a seasoned photographer, and it became apparent that McCallum intended to have him produce a stockpile of photos for EYOS's own use to post and distribute at will. That incident didn't make things any easier.

Because of other commitments, McCallum was also not making the tough monthlong Southern Ocean trip from Montevideo, Uruguay, to Cape Town. Instead, he asked if P. H. Nargeolet, Vescovo's consultant, could serve as expedition leader for the South Sandwich Trench dives—despite the fact that Nargeolet had no affiliation with EYOS. It was the second of the first five trips that McCallum would miss.

Vescovo had become so frustrated that he asked Geffen to recommend another expedition leader. Geffen passed along the name of someone he believed was qualified. When McCallum became aware of Vescovo's threat, he told Vescovo that he would be happy to recommend someone else, but warned that the list was very short.

McCallum finally contacted Lipton on January 20, detailing his issues with the agreement. He argued that the original NDA covered technical and corporate information, not social media or marketing, and not anything in the public domain. Further, he said that going public with the expedition at the media launch in New York rendered most of the NDA obsolete, except for any technical or trade secrets.

"The new Media Agreement goes a LOT further and we have no faith whatsoever that Atlantic will be fair or reasonable or that any agreement will be final," he emailed Lipton, concluding with a perfunctory: "We are considering the agreement."

Vescovo couldn't believe how hard McCallum was pushing back. "I can't help how he feels about Atlantic," Vescovo said. "He either has to work in the process, or he is saying he and EYOS cannot and they wish off the project. But there is no way I can let him play by a different set of rules when Triton, the science team, ship's crew, and all others are adhering to the same rules. He has to trust me, and the process I am overseeing, that they will get the media exposure and rights they seek. I remain incredulous over what, exactly, they are going to war over. His quip about Atlantic reveals how personal he has let this get. He is risking a very ignominious termination from the expedition."

A day later, Vescovo called McCallum, who was skiing. McCallum took off his skis, sat down on a snow drift and listened to Vescovo's reasoning. He then said that he would get back to him that afternoon.

McCallum phoned his partner, EYOS CEO Ben Lyons. "We are being bludgeoned into this and normally I would not sign this because I don't agree with it, but it's putting a lot of tension in our relationship with Victor," he said. "We need to make a decision."

The two discussed the best course of action. McCallum's feeling was that they would be selling out their morals if they signed it. He argued they didn't really need the Five Deeps job. His partner countered that they were helping make a historic mission happen and they had made a commitment to something that they believed in. McCallum reluctantly came around to his viewpoint.

"I didn't think Victor would fire me because if he had, he would have lost the crew," McCallum said. "If I had left, [Captain] Stu would have left. So, it's not like I felt threatened. It was only due to my absolute respect for Victor and what he was doing with the *LF*—that he had essentially given the world a fantastic tool—that we signed it."

As if things weren't touchy enough between McCallum and Vescovo, right in the middle of the Sturm und Drang over the supplemental media agreement, Vescovo learned that the permit for the upcoming South Sandwich Trench dive in the Southern Ocean had actually not been secured—an EYOS responsibility. The British government, which controlled the region, required the sub to be certified to a greater depth than the known deepest point of the South Sandwich Trench (approximately 7,300 meters), but the existing DNV-GL certification that Jonathan Struwe had issued was only to the 5,000 meters reached in the Bahamas with Struwe in the sub, not to the 8,376 meters reached off Puerto Rico.

McCallum had not told Vescovo—or had not known—that certification was required until days before the ship was set to sail on January 23. Not receiving the permit would plainly be a disaster, and any lengthy delay could cause the ship to miss the weather window and perhaps even delay the expedition an entire year.

Struwe saved the day. Though the certifier hadn't been in the sub on the Puerto Rico Trench dive, he had been on the ship in the control room. He agreed to provide a provisional certification to 8,376 meters. Based on Struwe's letter, two days before the ship was to set sail, the permit was approved.

That same day, both McCallum for EYOS, and Jones and Lahey for Triton, signed the restrictive media agreement. Still, Vescovo was concerned about McCallum carrying a chip on his shoulder as a result of feeling boxed in by Geffen, for whom McCallum carried a personal animus.

"He wrote several pointed but really threatening emails that probably did more damage than good," Vescovo later wrote to Lipton. "I am still turning over the idea of firing EYOS and getting a new expedition partner. He really pushed this far, far beyond where it needed to go and burned major bridge pylons with me." He continued, "And all this for what? So he can publish a few photos on social media that like, what, maybe a hundred people will see? It makes no rational sense. It must really just be a personal thing between him and Geffen. But I have zero time or patience for that. I'm trying to dive to the bottom of all five of the world's oceans, and we're jeopardizing that over social F-ing media. Full stop."

CHAPTER 13

HERE BE DRAGONS

The Southern Ocean is the least understood of the five oceans and the most unforgiving. Surrounding the frozen continent of Antarctica, which has no permanent residents, the ocean is dotted with massive icebergs and has the coldest waters, strongest currents, and roughest seas on the planet. Swells can roll for thousands of miles. Old mariners who risked their lives to carry supplies across the ocean used to call its latitude bands "the Roaring 40s, the Furious 50s and the Screaming 60s." The Five Deeps Expedition was headed into the heart of the 60s.

Mission number one for the expedition was finding and diving the deepest point in the Southern Ocean, which is generally defined as the body of water south of 60 degrees South latitude. This point was believed to be in the very southern portion of the South Sandwich Trench, a massive geological feature that straddles both the South Atlantic and Southern oceans. The Southern Ocean's deepest deep—which had never been identified or properly named—was believed to be located some 525 miles from South Georgia Island and 1,385 miles from the Falkland Islands, the very definition of a remote place.

Mission number two was conducing meaningful science, including mapping as much of the 600-mile trench as possible, as only very limited cartography existed. All data would be given to the Nippon Foundation's Seabed 2030 project, which aims to map the entire ocean by the start of the next decade.

Chief scientist Jamieson was on board, as was geologist Heather Stewart and Paul Yancey, of the Schmidt Ocean Institute. Jamieson and Stewart had done work in the Southern Ocean in 2018 and 2010 in the Atacama Trench off the coast of Chile, and the South Shetland Trench, but not in the South Sandwich Trench. Jamieson wanted to coordinate any new data discovered with their previous findings to advance understanding of the ocean.

"Now that everyone says the sub is working, there's a lot we can do," Jamieson said. "I'm hopeful that we can find a snail fish at Hadal depths, this being the only sub-zero temperature trench in the world, and to film what I believe will be densities of holothurians (sea cucumbers) at the deepest points. We'd also like to finally obtain a good map of the trench, as the ones we have are shit."

The *Pressure Drop* departed Montevideo, Uruguay, on January 24 with forty-four souls on board for a scheduled 28-day journey across the Southern Ocean to Cape Town, South Africa. The length of the trip and the anticipated brutal conditions were on everyone's mind. The day before the ship departed, one of the Triton crew even suffered a panic attack at the thought of being at sea for a month. After twenty-four hours of bed rest, he was cleared to sail.

The expedition was joined by a German ice pilot, Bernd Buchner, who was familiar with the iceberg formations that the ship would be dodging and the difficult seas caused by the ocean's powerful currents. Also on board was Dr. Glenn Singleman. Having a medical doctor was a maritime requirement on a journey this long, since the ship could be as far as a week or more from the closest port. An Australian with a specialty in remote medicine, Singleman was a veteran of James Cameron's expeditions, including the dive to the Challenger Deep, and an explorer in his own right.

Because of the distance and lack of access from any port, sub designer John Ramsay had also come on the trip in case something went wrong with the *Limiting Factor*. Even though the new manipulator arm would not be installed until after the Southern Ocean dive, the sub had been deemed ready to go with its core capabilities of manned descent and filming.

In the wake of the contentious debate about the supplemental media agreement, Vescovo printed out a notice prohibiting social media posts and taped it to the wall of the control room. At the first all-hands meeting, he explained that no photos or news of any kind about the trip could be posted in any forum, and he explained why. P. H. Nargeolet, who was serving as expedition leader in place of the absent McCallum, helpfully told the group this was actually normal for any deep marine expedition and related the code of silence that was placed on all of his dives over the years. "We need everyone to be team players on this," Vescovo said. There appeared to be grudging acceptance.

⁂

Because the conditions in the wild and unpredictable Southern Ocean, with its icebergs and swells that could stretch for miles, were expected to be the most challenging, before the trip there had been a lengthy assessment of the launch and recovery system. The problems with the LARS in the North Atlantic and during the dives in the Puerto Rico Trench had rekindled the debate over installing a man-rated A-frame which would allow the sub's occupants to board the sub and disembark while the sub was cradled on the ship.

Lahey was pushing hard to have the sub launched from the ship with the occupants inside. From his perspective, there was far less chance of damaging the sub as it was lowered from the ship. Captain Buckle was uncomfortable with that plan, because it was against ship regulations for him to launch a person from a non-man-rated crane and he could possibly put his captain's license in jeopardy for doing so. The only provision for the current A-frame being operated with the pilots inside was if an emergency recovery was necessary to prevent potential, serious bodily harm to them.

A man-rated crane would cost in the neighborhood of $750,000 fully installed, about half of the original estimate. Nevertheless, the biggest concern was not money, but *time*. If the expedition had installed the crane after Puerto Rico, it would have pushed the Southern Ocean dive to the following year as the weather window would have been missed, thus adding nine months to the dive program.

During the discussions, a breakthrough idea came from DNV-GL's Struwe, who found and approved a workaround of sorts. He reasoned that as long as the sub was suspended off the back of the ship and partially in the water, it could be launched with the pilots inside. Lahey came up with the idea of placing a horseshoe-shaped bumper assembly at the stern that would allow the LARS team to lock the *Limiting Factor* in place for loading and unloading. To do this, the sub would be lowered to the point that it was just touching the water, and then a fiberglass-like material gangplank with a nonskid surface would be extended for the pilots to board the sub from the ship. It was not a perfect solution, but it was dramatically better, and safer, than the dicey launches the team had gone through on the open seas.

"The plank's a game changer," Lahey said. "Now everything is tied in so the ship and the submarine will be moving together rather than at odds."

After conversations among Vescovo, Buckle, McCallum, and the Triton team, several additional new features and steps to the LARS process were instituted to improve safety, reduce risk, and enhance efficiency. These included the use of a lightweight titanium hook with a simple spring-loaded hook closure and the addition of a soft master link that would be both lighter and more manageable than the steel master link currently being used. This would make it easier for the swimmer to attach and detach. The team also experimented with several new sea anchors to both pull the sub away from the stern of the ship and to make it move much more in a straight line when behind the ship, and not in a zig-zag fashion as it was prone to do. Buckle was initially concerned about the "walk the plank" solution. However, with Struwe's blessing, he okayed the process.

With this solution in the offing, Vescovo decided to stick with the current A-frame. If the expedition could make it through the Southern Ocean dives with this new approach, he reasoned that they would not need a man-rated crane for the final three ocean dives, as conditions in the Indian, Pacific, and Arctic would almost certainly not be as rough.

∽

The expedition team spent the first few days at sea marveling at the fascinating marine life they witnessed as they made their way south, assessing the optimal time to dive, and readying the *Limiting Factor* for its cold water debut. The conditions were such that they might only allow for a single solo attempt at the bottom, making preparation even more critical.

The sailing days were long and cold, but seeing dolphins and humpback whales breach the surface against the backdrop of icebergs was a once-in-a-lifetime experience for most. At one point when the ship pulled in close to an island, there were a flock of chinstrap penguins frolicking about. Even with all the pressure on the team, there was a feeling that the trip would enhance everyone's life in some small way, giving them memories that few people have.

"That is what an expedition like this is really about, experiencing things you cannot see elsewhere," Singleman said. "The thing about the Southern Ocean is that it is alive. We as a species have polluted so much of the planet, but we haven't done that completely to the Southern Ocean because its location hasn't allowed us. To stand on deck and look at the world with a new perspective allows you to see a raw power that this wilderness provides."

On the practical side, the expedition team kept a very close eye on the weather forecast. Over the course of the first week, the ocean was at sea state 2 to 3, rough but tolerable. Far ahead of them, near the South Sandwich Trench, were front after front barreling across the ocean like freight trains.

At daybreak on January 31 after a week at sea, the ship tucked into the Cumberland Bay near South Georgia Island. After clearing customs, Triton wanted to do a short dive so the crew could adjust to the new launch and recovery system using the boarding ramp. In the near-perfect calm of the inlet, the dive went well and provided a great shot of confidence to the team.

After thoroughly cleaning all clothing and footwear that would be worn on South Georgia Island, the group took the tender in relays to the island to pay homage to Sir Ernest Shackleton, who was buried in Grytviken, an abandoned whaling town on the island. The cleansing was to prevent any bio-contamination, as the island was heavily populated with fur seals and king penguins.

Everyone gathered around the grave with a bottle of whiskey, a tradition for visitors to the great explorer's gravesite. Vescovo offered a toast to Shackleton, who died at forty-seven and whose birthday was five days after Vescovo's own.

"Sir Shackleton was a great explorer," he said. "He is best known for the expedition when his ship got trapped in ice and over multiple years was able to keep his crew intact, and then in a small boat sailed to South Georgia Island and arrange for a rescue of his crew. No one died. He was a great leader."

The team toasted. "To exploration," Vescovo added. And then, as tradition dictates, everyone tossed the remaining whiskey in their glasses onto the grave.

❧

When the *Pressure Drop* crossed the Antarctic Convergence zone, the area circling the continent, there was a sharp drop in the air and water temperatures. As the ship passed the volcanic Thule and Cook Islands, the southernmost in the South Sandwich Islands chain, icebergs began appearing on the horizon. The ship was being vigorously rocked, sending lighter items flying and making it difficult to sleep. Many took to placing a U-shaped life preserver under their sheets to prevent them from rolling off.

Though the trip had only reached the ten-day point, the remoteness was setting in for those not accustomed to being at sea for long periods of time. "A combination of the rough seas, the lack of activity, and sheer feeling that we were in the middle of nowhere started to wear on many of us," explained Richard Varcoe, the website operator. "And the end was so far off."

The ship explored the extreme southern portion of South Sandwich Trench while waiting for a weather window to dive. The maps of the trench created from the sonar data revealed in detail three major depressions south of 60 degrees latitude, the part in the Southern

Ocean that had not been previously mapped with high-resolution equipment. Most of the world's ocean maps are filled in with satellite altimetry—estimates of seafloor topography based on gravitational changes of the sea surface. This is extremely low-resolution mapping that encompasses the whole globe, but because each pixel is around 500 meters x 500 meters, the features below could be hundreds or even thousands of feet shallower or deeper. The new bathymetry maps were brilliantly revealing undulating mountain ranges and their associated canyons on the bottom of the ocean. The mapping expert, Cassie Bongiovanni, also located the deepest point in the area for the dive, with a calculated depth of 7,432 meters (24,386 feet).

"The three basins were already known," Bongiovanni said. "What was special about this one was the deepest point was in the one no one expected it to be."

As the calendar turned to February, a demarcation line not felt in the least on a ship at sea, Buckle and Buchner, the ice pilot, determined that there was a two-day diveable weather window, with the most promising day being February 3. With a dive on the docket, the energy level picked up.

On the morning of the 3rd, the sun's rays pierced the threatening clouds, but the sea was ignoring the sun, kicking up a fuss. The sky notwithstanding, the conditions to launch any open water operation were predictably unkind, with swells dominating the area and a partnering wind kicking up the waters to sea state 3. This was at the limits of what they deemed to be safe, but still doable.

The Triton team forged ahead, preparing for Vescovo's solo dive. Even as the winds continued to whip across the ship, all in all everyone felt it was diveable. An obvious hazard was the water temperature, which was below freezing (though salt water doesn't actually freeze). This posed a danger for the swimmer who had to "ride" the sub to attach the drogue and to detach and later reattach the tow line.

On the pre-dive check, there was a fault in the Toughbooks, the panels (or graphic user interfaces, GUIs, as they were usually called) used for data readout. The Ethernet cables attaching them to the submarine's data bus were not properly connected. Blades was able to debug the problem in thirty minutes.

As Vescovo prepared to board the sub, Lahey, who was operating the A-frame, yelled down: "The moment is now . . . have a great dive."

Vescovo came back with, "It'll be awesome."

The revised pilot's entrance to the sub from the gangplank made things dramatically more safe and efficient. The tow time was minimal, and in a matter of fifteen minutes, the *Limiting Factor* was released. Vescovo reported that he had a green board of electronics. Lahey, in the comms room, cleared him to dive.

"And down we go," Vescovo said.

On the descent, Vescovo could feel the cold coming in through the metal bottom of the sub, but he was prepared, wearing the same UGG booties he wore on Mount Everest. Looking out through the viewports, he could see schools of jellyfish and plankton swimming around. He was surprised how dense they were in such cold water—a veritable feast for the whales and penguins that migrated here each year. He kept thinking, you don't have to be a scientist to be awed by this view.

All systems on the sub were working fine. Initially, Vescovo was seeing condensation around the bottom viewport, about a tablespoon of water every few minutes. He monitored the ingress to determine if it were necessary to abort, but around 2,000 meters, the condensation stopped. It was just water vapor coming out of the air in the capsule due to the cold temperature.

After passing through 3,000 meters (9,842 feet), he heard a low humming noise that sounded like a lawn mower in the distance coming from inside the sub. After checking various electrical systems, he couldn't figure out what it was. "I have a feeling all is not right," he said to himself.

By the time he reached 3,200 meters, his communications with the *Pressure Drop* and the *Learned Response* on the surface suddenly failed. He could not hear Lahey, who was in the control room conducting the quarter-hour checks.

Vescovo noticed that he was descending at a rapid rate and thought that might be the reason for the comms failure. To arrest his descent, he blasted the thrusters at full power for twenty seconds. When he finished the test, he had dramatically higher noise levels in his headset, around ten times as much. It was so loud on the starboard side headset (the pilot sits on the port side) that he could clearly hear it from across the interior of the capsule on the unused headset.

Sensing an issue, he tried to contact the surface repeatedly, but had no success. He also tried texting, but nothing came back. There was only silence. Silence at such a depth has a hard quiet.

In the comms room on the ship, Lahey grew concerned after a second missed quarter-hour comms checks.

"*LF*, comms check . . . comms check," Lahey barked into the microphone.

There was no reply.

"What the fuck is going on?" Lahey said to Blades, who was sitting next to him at the control board. "*Response*," Lahey said addressing the tender that also monitored communications, "did you hear comms check?"

"Negative . . ." came the reply from Hector Salvador, who was piloting the boat.

A hush fell over the comms room. The silence was soon broken with Vescovo's voice coming across, very faintly, acknowledging that life support was good.

"All stop!" Lahey yelled. "Roger that *LF*, I have you loud and clear . . ."

Vescovo's garbled voice continued, "Present depth 3,331 meters . . . I cannot hear you."

Lahey turned to Blades. "He didn't hear me. . . . he didn't acknowledge," Lahey said. Then, he tried again. "Do . . . You . . . Read . . . Me . . ."

A minute later, Vescovo's voice was heard again. "Surface, surface, how do you read?"

"Loud and clear," Lahey said. "Status!? Status?"

Again, there was no reply.

Lahey hunched over the comms panel. "What the fuck is going on where he can't hear us?" Lahey said.

Blades continued to adjust the sound levels in the effort to reestablish communications.

"What the fuck, Tom?" Lahey said in a plaintively desperate tone.

The soft-spoken Blades, who was working through the failure and finding solutions in his head that made sense, finally replied, brushing Lahey off with an "It's interesting . . ."

In the sub, Vescovo was trying to figure out what was wrong. He reset the software of both the port and starboard electrical "sides," but there was no change. He rebooted the starboard Toughbook and then relaunched the software, also to no effect. He did not try this on the port side, as he didn't want to risk losing his primary software.

Next, he cycled the power in all nonessential systems in hopes that might reboot the system because the electronics were all interconnected. When he turned off the altimeter, the noise in his headset almost doubled in volume. He noticed that no other switch except the altimeter had the effect of increasing the noise level. Because the sub was otherwise fully operational with respect to thrusters and its power levels, he considered it too dangerous to cycle any of the batteries for fear of them not coming back online.

Nothing he did brought the comms with the ship back online. He executed standard underwater phone check-ins two more times, but with no reply and unusually high background noise, he suspended the calls. He was certain that all aspects of the comms system had failed and that the surface was not receiving any word at all.

After his troubleshooting failed to fix the issue, he began to consider his options. He knew that Lahey and the team would be upset if he did not abort, which was the standard procedure in a total loss of comms, but the sub was otherwise functioning fine. He checked the electronics, and the oxygen and CO_2 readings. Everything was perfectly in order. He reasoned that the cold water or some electronic gremlin must be causing the problem.

For Vescovo, it was decision time. He was about 3,000 meters from the bottom, where no human had ever been. He decided that he had come all this way and was not going to abort for a failed comms check, even though that was the agreed-upon procedure. There was no guarantee he would even have a good weather window another year from now. He felt he was oh so close.

He said to himself: "I don't want to come back here. I've got 'only' 3,000 more meters to go and I've been down deeper. So I'm going for it."

Nearly an hour had now passed without any communication between the ship and the sub. Vescovo had stopped checking in.

Lahey was beside himself. Without being able to talk to the control room, Vescovo wouldn't know where the ship was when he ascended. He was so deep that there was also a small risk of an iceberg drifting over him during his dive, yet he would have no way of knowing. Lahey was also irritated that Vescovo was not observing the quarter-hour comms checks, as he assumed Vescovo had concluded all comms were out.

"Fuck . . . fuck . . ." Lahey kept muttering under his breath.

Meanwhile in the sub, Vescovo continued descending. "Well, now I'm really alone," he said out loud. "I don't have comms with the surface. I'm in a submersible by myself going down to the bottom of the Southern Ocean, where no human has ever gone."

The irony was not lost on him. He was a loner by nature who climbed mountains and now dived a sub to escape from everyday life. He certainly couldn't get further away than he was now. He was one man doing one thing with no one else there. The outside world didn't exist. It was the epitome of what he had wanted to do—diving the bottom of the oceans, alone and unafraid, relying completely on himself. Heck, even spacecraft have copilots.

It was the ultimate in solitude. To enjoy and reflect on the moment, he played the song "Deeper" by the Fixx. He figured it was appropriate.

Then he turned on the altimeter. The readout showed that he was 150 meters from the bottom. His excitement grew. With about 30 meters to go, he tested the VBT weights that would make the sub neutrally buoyant and allow it to hover just over the bottom. These had given the team fits on previous dives. He pulled the trigger and heard the right sounds. "Ker-chunk, ker-chunk"—they both dropped.

Just in case something had changed with the comms, he checked in with the ship "in the blind."

"I'm about five meters from the bottom," he said into the abyss, as a pilot would do flying into space.

Then he touched. "Surface, *LF*, at bottom. Repeat, at bottom."

The comms room heard him. A cheer—and a sigh of relief—went up.

He peered out the viewport. The sediment was dense and coarse and not as peanut butter–like sandy as the bottom of the Atlantic Ocean. He noted that although Stewart, the geologist, said that he would see lava rocks, he did not.

He decided to try to find a lander. However, without any direction from the ship, he didn't have even a rough idea where the lander was. He checked the sub's limited range sonar, but didn't see any sign of it. The sonar had a maximum range of 200 meters, but a true effective range of about 100 meters. Employing a circular pattern, he started the search. As he circled, at least he would be filming the bottom with the sub's high-definition exterior cameras.

Then it became a question of how long he would stay down. He reasoned, "The longer I stay down if they think I'm dead, the more I'm torturing them." He decided on a compromise of one hour.

On the ship, after Vescovo's report came through that he was on bottom, Blades turned to Lahey. "I'll bet Victor stays down there an hour and comes back up," he said.

That was exactly what Vescovo did.

After an hour of driving around the seafloor, he said, "I think that's a wrap." He toggled the switch to release the surfacing weight to begin the ascent. Hearing it drop washed away all the tension. He put on the Deep Purple song "Smoke on the Water," sat back and relaxed.

Up he went, though without comms he wasn't sure where he was in relation to anything above him, namely the ship or an iceberg. As a precaution, he decided to ascend as slowly as possible in the event that he bumped into something. If he did, he would just head off on a vector for about fifty meters and try again.

Over the underwater telephone (UWT) on the way up, the comms went in and out briefly. He heard the distinct, Spanish-accented voice of Salvador from the *Learned Response*. He could barely make out what Salvador was saying, which sounded like "Congratulations." He had enough sporadic communications with the surface to get the sense that he was cleared to surface—that somehow they had a fix on him. Still, he ascended extremely slowly. He eventually floated gently to the surface, whereupon he dropped his freeboard weight to put the sub as high in the water as possible.

The ascent had taken just over two and a half hours. When Vescovo surfaced, he switched to his VHF radio and reported that he was on the surface. However, there was no response. He was puzzled because the VHF radio was not connected to the comms system that had failed. He sat there, wondering where he was. Because of the very sporadic comms on the way up, he had no idea if the *Pressure Drop* was minutes or hours away.

As he bobbed on the surface, a thought popped into his head: how long do I wait before opening the hatch and using the Iridium phone to literally call them up and give them my coordinates? The phone would only work if the hatch was open, but if a big wave arrived at the same time, it could swamp the capsule. Plus, he would be subjected to the freezing water of the Southern Ocean with no way to warm himself up. He decided to wait at least four hours before even considering that option. Only a mountain climber could think in these terms.

The situation was even more unnerving for him than not being able to hear the control room on the way down, because when he surfaced, he expected the VHF radio to work. They weren't *that* far away, were they, he thought? But, who knew? How could he? Moments make up life, and Vescovo was used to having some control over them, or at least relying on science and math, his bellwether trusts, his certainties, his buoys, to define them. He had, but they offered no relief now. Oddly, with all the know-how, ingenuity, and revolutionary physics that had gone into this project, he was just a man in small craft bobbing in icy water, waiting for someone to come and get him.

Fifteen minutes later, an eternity inside the titanium sphere with no proactive moves to be made, he heard a crackle over the radio, followed by the sub being boarded by the swimmer.

Hooking the sub up to the ship in the cold waters was a struggle for the swimmer, Steve Chappell, who was wearing a dry suit to protect himself against the polar waters. At one point, he took off a glove to get a better grip on the hook from the main tow line. The freezing conditions assaulted his hand, and it instantly went numb. Struggling, he manhandled the equipment into behaving, and eventually, the sub was hoisted partially out of the water and the hatch cleared for opening.

Again, there was no precedent for any of this, leaving the team feeling unrestricted.

Vescovo emerged from the *Limiting Factor* holding up two fingers signifying two oceans down. "Great job, team!" he said.

It was so cold that by the time he walked the gangplank to the ship everyone except the core recovery team had gone back inside. He passed Lahey on his way in. "Welcome back," Lahey said, adding, "We'll talk about procedure of the comms checks later . . ."

Vescovo nodded. He was hungry from the dive so he headed to the galley after changing into warmer clothes. The cook prepared him a big bowl of his traditional post-dive meal, spaghetti Bolognese, and he sat alone and ate, mentally decompressing.

Once the sub was back on the ship, Chappell was taken by Dr. Singleman to be treated for hypothermia. Shivering and nearly purple, he was placed next to the heaters in the sub's hangar. Fifteen minutes later, he regained feeling in his hand and his body temperature returned to normal.

"I felt the stinging of a thousand needles on my fingertips, and all I thought was that I have to get the sub hooked up and see this through," Chappell said of the struggle.

That night when the landers were recovered, conditions were easily the roughest they had been on the trip. The ocean had reached sea state 4. It was pitch black and snowing sideways.

Post-dive, there was good news and bad news. On the good news side of the ledger, the three deployed landers—one of the expedition's, *Closp,* and the other two smaller ones belonging to Jamieson—had all been recovered. And even though Vescovo hadn't seen the landers, *Closp* had fleetingly captured the shining lights of the *Limiting Factor* touching bottom. He had been so close to finding it when executing his search pattern.

For his part, Jamieson was excited by the finds from the lander. The cameras had captured far larger than expected densities of the sea cucumbers (holothurians), which Vescovo had also seen through the viewports. In all, the landers had picked up four new species of snail fish closely resembling those from Jamieson's previous findings in the Southern Ocean's Atacama Trench, and a new species of grenadier that he had only seen before in the neighboring South Shetland Trench. The promise of more science in the days ahead remained.

The bad news was that when Vescovo cycled the power while trying to troubleshoot the comms issue, he had shut off the sub's external high-definition cameras, and they never came back on. This meant that there was no footage from the hour he spent exploring the bottom, save for the lander's footage. For whatever reason, he also forgot to film the bottom with his phone camera through the viewports as a backup, as he had done previously.

Lahey was upset that Vescovo had not continued to check in and had not aborted when communications went out. "That's procedure and you need to follow it," he told him. "We could hear you."

Vescovo apologized for not continuing to check in even though he could not hear the control room. "I was totally convinced you couldn't hear me," he said. "Heck, I couldn't even hear myself when I transmitted because of the interference. But I'll continue to make standard reports even if I think comms are all shot. And if I have a good sub and good life support, I will keep going down. I don't consider comms to be a mission-kill criteria if everything else is good. Maybe for normal sub operations like most commercial dives, but not where we are operating."

Lahey decided to table the conversation for another time. "Let's leave it at the comms were a problem, but the mission was a success," he said.

Blades had analyzed what went wrong with the comms, and it was what Vescovo had suspected during the dive. The water temperature outside the sub was minus 1.6 C (29 degrees F), and it was 4 degrees C (39 degrees F) inside the sub. The cold temperatures caused the bands from the communications to spread out, leaving Vescovo hearing only noise on his end. Lahey would solve the problem by replacing the filtering circuit so that would not be an issue again. As for the VHF radio, Vescovo had surfaced too far from the ship, and his small hand-held radio didn't have the power to reach the ship many miles away. To correct that, Vescovo requested that a more powerful radio be installed for future dives.

The final depth of the dive was recorded at 7,433.6 meters (24,388 feet) in a place that has not been previously measured or named. Because Vescovo was the first to reach it, he would be allowed to name it. He decided that it would the Factorian Deep, after the vessel that brought him there and back.

ℐ

The *Pressure Drop* then continued mapping the South Sandwich Trench en route to the Meteor Deep, the second deepest point in the Atlantic Ocean, for a science dive with Vescovo and Jamieson, weather permitting.

On February 5, three landers were deployed in the middle of the trench, one with a core sampler. Stewart and Jamieson had received an £80,000 (just over $100,000) grant that included the retrieval of a sediment sample and the analysis of it.

The day turned into a series of disasters on the science front. The lander with the core sampler malfunctioned and did not pick up any sediment. One of Jamieson's landers didn't surface at all. Similar to the situation with the lost lander at sea trials, no one was quite certain why.

Buckle calculated where it should be based on the current. He called his crew to the bridge with binoculars and searchlights. Three hours of searching produced no sign of the lander.

The second lander surfaced later that night in very rough conditions. As Jamieson and the crew were recovering it, the lander was swept up in high seas and the ship's propeller sliced the rope attaching it to the crane. The lander plummeted to the bottom of the ocean.

Jamieson couldn't believe it. In a very short period of time, he had lost his sediment sample and two of his own $100,000 landers. Vescovo proposed a name for this site: The Bitter Deep.

Things continued downhill for Jamieson. The planned science dive in the Meteor Deep was canceled because the weather bordered on sea state 4 and didn't look to improve. The decision was made to head east for Cape Town and continue mapping the trench. Jamieson was in the darkest mood anyone had ever seen.

But about thirty-six hours later, Buckle and Buchner noticed the weather at the newly revised position of the Meteor Deep was turning to a calmer sea state 2, imminently diveable conditions especially with the new gangplank boarding system. The decision was made to literally turn the ship around and return to the site for a dive. After the long crossing and the prospect of heading for land, the move taxed several of the team mentally. No matter, the mission came first.

On the morning of February 10, which happened to be Vescovo's birthday, the Triton crew prepared for the science dive. Conditions were rough, with large swells. One lander was deployed early that morning. The sub preparations took an extra forty minutes, but the dive was still a go.

Vescovo and a very excited Jamieson boarded the *Limiting Factor*. It was lowered partially into the water. Just as the team was ready to launch, Buckle informed Lahey that the delay had caused the ship to pass the drop point where the lander was by some 400 meters. Lahey told Buckle to turn the ship around to return to the drop point. However, once the ship reversed course, the waves were coming from the rear instead of the front, known as "a following sea." Though this was a dicey proposition with the sub so close to the ship, none of the launch crew raised any red flags. It was a serious mistake.

As the ship was returning to the drop point, a powerful wave raised the ship. The sub, acting like a pendulum, went underneath the foam barrier, pushed the barrier up and out of way, and the sub impacted the ship. The collision sheared off one of the standard definition cameras and cut the cable attached to it.

Inside the sub, Jamieson looked at Vescovo, eyes wide with surprise and no small amount of anxiety. Vescovo assured him it was okay. "We still have a green board," he said.

The *Xeno* was next to the sub. The crew aboard the tender looked at the sub and radioed Lahey that there didn't appear to be any significant damage, so Lahey granted permission to dive. In the rough waters, the team on the *Xeno* saw that the camera had been sheared off—but that was not mission-critical. However, they had apparently not seen that a key control cable had been cut as well.

As the dive commenced and the sub descended rapidly through the water, one of the crew watching from the upper deck spotted a small oil slick over the sub's launch point and radioed Lahey, who was manning the comms system. It was quickly determined that the oil was from the inside of the cable running to the camera that had been broken off in the collision with the ship. The danger was that eventually all the oil would drain from the electrical system and allow sea water to flow into the junction box, which would short it out. In addition to losing a $100,000 junction box, if that occurred, the freeboard weight would release, and the sub would be too light to make it to the bottom.

With the sub at 400 meters, Lahey radioed Salvador on the *Learned Response*. "Instruct *LF* to abort the dive," he said. "We have an open hose where the camera was damaged." Salvador alerted Vescovo, requesting him to abort.

Vescovo didn't hesitate. "Roger that, aborting dive," he said.

On the way up, Vescovo, in a clearly frustrated tone, asked what had happened, and was told about the oil leak.

Lahey turned around to the group gathered in the control room. "There are times when you have to call it and it's unpopular, but you have to do it because it's safe," he said as he prepared to help the LARS team on deck.

In that short period of time, the seas had grown significantly rougher, and a wet snow had begun falling. The Southern Ocean was baring its teeth. Later, Vescovo remarked that it felt like the ocean was now taunting them and it was time to get the hell out of there.

After the sub surfaced, the LARS team struggled to get the *Limiting Factor* in position to bring it back on board. When the sub was close to the ship, the swimmer boarded it and attached the lift lines—but one of them snapped in the rough seas, causing the sub to turn fully around.

Reattaching the cable would be nearly impossible in the cold, windy conditions. Before waiting for the situation to become dangerous, Lahey radioed Buckle that they needed to implement the emergency procedure and lift the sub directly on board with the occupants inside. The captain agreed.

As the LARS team was recovering the *Limiting Factor*, it smashed into the stern of the ship and crushed three of the thrusters. Minutes later, the sub was hoisted out of the water and loaded onto its cradle backward. After the water was pumped out of the trunking, Vescovo and Jamieson exited disappointed, but without issue.

∞

To avoid more severe weather that was on the way, the *Pressure Drop* immediately headed off the dive site for the eleven-day transit to Cape Town. Jamieson had had a bad several weeks, losing his landers, not getting the core sample, and having his first dive aborted.

Lahey took responsibility for the mishap that occurred when the sub was launched in a following sea on the science dive. "In hindsight, I should have objected to continuing the launch process without first passing the site and returning on a course into the weather," he said. "This was my fault and I take full responsibility for the outcome."

It was yet another lesson moving forward.

Over the next few days, blue skies and calmer seas prevailed. En route to Cape Town, Jamieson asked Vescovo if he could survey and deploy a lander in the Agulhas Fracture Zone, to determine if it qualified as being in the Hadal Zone (below 6,000 meters). Vescovo, feeling guilty about the science setbacks on the trip already, readily agreed.

The team mapped the length of the Agulhas Fracture Zone, with the ship stopping at the anticipated deepest point. The sonar recorded the maximum depth to be ~5,400 meters, some 1,000 meters shallower than expected based on previous reports and global bathymetric estimates from the General Bathymetric Chart of the Ocean, a publicly available bathymetric chart. The shallower depth readings were particularly disappointing to Jamieson as the Agulhas Fracture Zone could no longer be classified as part of the Hadal Zone. However, this is not uncommon in ocean exploration and highlights how little we know about the depths of the seafloor. Despite this, a lander was still deployed to the deepest area, where it stayed overnight to record data that could be spliced into a bigger-picture story at a later time.

Overall, Jamieson remained disappointed with the science aspect of the expedition and had become upset with the film crew. "I wasn't enjoying being around the film crew trying to turn my extraordinary bad luck into entertainment, but then out of nowhere I was offered my first dive that after a massive buildup ended up aborted at 400 m which gave the film crew more and more reasons to try and humiliate me to the point I stopped playing ball," he said, summing up the experience in his typically direct manner.

Jamieson had spent eighty-eight days on the ship since the beginning of the FDE and felt he had little to show for it. His position was that the science had been getting short shrift from day one. His week in the Puerto Rico Trench for science was killed. The loss of the manipulator arm handcuffed the gathering of sediment samples in the Puerto Rico Trench or in the Southern Ocean. Even the big victories weren't scientifically relevant.

"Victor dived to the deepest point of the Puerto Rico Trench, crashes into the seafloor, goes in circles and doesn't bring back a sediment sample," Jamieson said. "That's utterly useless from a scientific point of view."

The results of the Southern Ocean trip were only slightly better, but the setbacks were getting to Jamieson. "The start of this trip was amazing, and it felt like we were going somewhere on an expedition, and not just pissing about doing more tests on the sub," he said.

The lander data was somewhat useful when bundled with the previous work Jamieson had done and the trap samples revealed four new species of shellfish. The mapping of the trenches was the real pathbreaking winner.

"But everyone talks about the sub, and again it produced nothing of value to me," Jamieson said. "As far as I was concerned I had lost way too much time—and now all my gear—to this expedition and it doesn't look like it is going anywhere but a few token hero dives. That's fine, as it was never the core mission of the FDE, but as chief scientist, science is kind of important, so I have decided to respectfully bail."

Buckle was supportive of Jamieson and echoed his disappointment. "Our frustration is amplified because when we first got involved, there was more time on location for science," the captain later explained. "But that time was all clawed back because the sub wasn't ready and hasn't worked properly."

On February 23, the *Pressure Drop* ported in Cape Town. While the core mission—to dive to the deepest point—was achieved, where there was success, there be dragons. The dragons, it seemed, had caused the chief scientist to call it quits not even halfway into the Five Deeps Expedition.

Vescovo decided to give Jamieson space to regroup. He felt that the *Limiting Factor* was just getting started as a science tool, and with the installation of a new manipulator arm would reach its stride. "I remain convinced that with each dive evolution, we will continue to learn and perfect this system to the point where by this point next year, it will be truly formidable, eminently predictable, and a great addition to the scientific arsenal for deep-sea exploration," he said. Time, perhaps quite a bit of it, would tell if he was accurate.

CHAPTER 14

THE JAVA SCRIPT

The Indian Ocean is the warmest ocean on the planet, the least fertile for marine life, and the most polluted with human contamination. An extreme magnification came in 2019 when scientists made headlines by finding some 414 *million* plastic items, weighing 238 tons on a remote archipelago that had piled up because of a total lack of a clean-up effort. Even beaches in the toniest areas of tourist-heavy Bali that are swept twice daily can come to resemble dumping grounds when high tide arrives.

The ocean's Hadal Zones in the Diamantina Fracture Zone and the Java Trench remain largely unexplored and are believed to be a settling ground for the human contamination. When the Five Deeps Expedition planned its March 2018 dive, there wasn't verifiable data on which trench contained the deepest point. And there was scant examination of the area where Earth's tectonic plates had shifted in the Java Trench in 2004 in a massive undersea earthquake, causing a tsunami that resulted in massive destruction and 250,000 deaths in Indonesia. And, of course, no human had been to the bottom of either deep. Exploration had potentially untold benefits.

En route to the Indian Ocean, the *Pressure Drop* sailed from Cape Town with its transit crew to Freemantle, Australia, where it picked up the Triton team to install the new hydraulic manipulator arm on the *Limiting Factor.* The titanium frangibolt securing the arm to the sub had snapped in the Puerto Rico Trench, resulting in the arm falling off and being left at the bottom of the Atlantic. Vescovo had refused to pay the $300,000 sticker price for a new arm, citing a fundamental design flaw and maybe a lack of maintenance procedures to catch its weakening over time. Triton went back to Kraft TeleRobotics, the manufacturer, and negotiated a far lower price and in compensation for Triton's faulty setup, paid for the arm itself. The frangibolt on the new arm had been remounted so that it did not have the degree of lateral stress and also so that the stresses would be absorbed by the bracket around the bolt rather than the bolt itself.

On the way to Freemantle, the hydrographers mapped the Diamantina Fracture Zone, located 700 miles west southwest of Freemantle, to determine if there was a deeper point than the Java Trench. (A fracture zone is a deep point in the ocean where both sides are part of the same tectonic plate, as opposed to a trench, where two plates meet.)

Jamieson had built a makeshift lander to help verify the depth of the Diamantina Fracture Zone and confirm that it was not deeper than the Java Trench, which was believed to have the deepest point in the Indian Ocean. Years earlier, when Sir Richard Branson was planning a five deeps mission, which he labeled the "Five Dives," his team had identified the Diamantina as being slightly deeper than the Java Trench. The question would be settled by the combination of the Kongsberg EM-124 sonar on the *Pressure Drop* and a physical lander measurement.

Jamieson left Erlend Currie, a young crew officer, in charge of the makeshift lander he built, and thus dubbed it the "Erlander." Currie operated the lander with the help of a cheat sheet and some extensive WhatsApp conversations with Jamieson, who was sitting on his sofa at home in Scotland contemplating whether or not to return to the expedition. The lander captured several amphipods, making Currie—a sailor by trade—the first person to collect biological samples from the Diamantina.

The official depth measurement came in at 7,019 meters (23,028 feet), about 150 meters less than the Java Trench data available. Still, the *Pressure Drop* would map the entire Java Trench to certify the deepest point, and also to complete the first ever detailed bathymetric map of the trench. The lander also managed to bring back the only sediment samples ever taken in the Diamantina.

During the break after the Southern Ocean trip, Jamieson thought long and hard about rejoining the expedition. He had left all of his science gear on the ship. If he quit, he reasoned that he would have to fly to Bali on his own dime to pick everything up. Begrudgingly, he decided to continue for the Java Trench dives, in part because the area was largely a scientific mystery. He hoped that two crucial aspects could be explored, the effects of dumped plastics on marine life and the trench's tectonic plates.

"The science is fuck all on this expedition, but I'm going to give it another go," he bluntly put it.

Indonesia is home to majestic rainforests with active volcanoes and breathtaking waterfalls, the dreamy beaches of Bali and the spiritual center of Ubud. It also turned out to be home to a byzantine governmental bureaucracy that makes the one in the U.S. seem straightforward by comparison.

In the days leading up to the Java Trench dives, the scramble for a science permit continued. Rob McCallum and Andy Shorten, a maritime expedition expert in Bali who had been hired to help, had been jumping through hoops for months to secure the permit. Vescovo himself had flown to Jakarta in late December on two days' notice and given a presentation to some thirty government officials at the annual Indonesian Oceanographic Community conference, during which he invited Indonesian scientists onboard and even offered to take one of them on a dive, schedule permitting.

"Getting Java Trench permissions is shaping up to be a complete mission all on its own," a frustrated McCallum emailed Vescovo. He had been working on the permits for nine months, and things were coming down to the wire.

As the chief scientist of what was classified as a research vessel, Jamieson was the permit's formal applicant. He too had spent weeks corresponding with Nugroho D. Hananto, the deputy director for Data

Management and Dissemination for the Research Center for Oceanography at the Indonesian Institute of Sciences (LIPI), who went by Nug for short—or as Jamieson took to calling him in internal correspondence "Nuggie."

Nug was tasked with gathering the proper information so that the higher-ups at LIPI could sign off on the permit. However, the entire process was bogged down in red tape. Even with the invitation extended for Indonesian scientists to come on the trip, things were going nowhere. With twelve government ministries and the Indonesian military needed to sign off on the expedition in some way, and with an upcoming presidential election, there was speculation that no one in the government wanted to sign off for fear that something would somehow go wrong and reflect negatively on the current government.

The day the expedition team and the film crew boarded the *Pressure Drop* in Denpasar, Bali, Shorten received a phone call from a government contact telling him that the permit would not come through. McCallum checked the application status on the government's website. Under the heading "Decision," it said "Rejected." Under the next heading "Remarks," it was blank.

The expedition was now in a bit of a gray area. Under the International Law of the Sea, the ship could sail in the Exclusive Economic Zone (EEZ) of Indonesia—defined as the area between 12 and 200 nautical miles off its shore—and dive the sub to test its equipment, but not conduct "marine research." One major problem with the Law of the Sea is that it doesn't precisely define what "marine research" is—it has been debated for decades. Furthermore, the law stipulates that its signatories (such as Indonesia) are required to support marine research unless there are specific, valid reasons not to. Without providing one of the approved, specific reasons for the permit's denial, it appeared that Indonesia itself was in violation of the law.

What was clear was that the Indonesian government did not have the right to stop a ship from sailing, or even having a submarine dive in its EEZ. However, without a marine scientific permit the expedition was prevented from doing what would be obvious "marine research," such as removing any sediment samples or marine life. Simply put, "visiting and filming" and "testing equipment" appeared allowed under the Law of the Sea, provided nothing was taken.

Beyond limiting the scope of the science, there were practical problems. Though the Indonesian government had no real right to do so, there was a slight fear that its navy would shadow the ship with a patrol boat and monitor its activities and request it leave the country's EEZ. Even if that didn't happen, returning to port in Bali, where the expedition team and film crew had flown in and boarded, was deemed too risky, as the ship and its crew could be detained once they were back in Indonesian territorial waters (twelve nautical miles from shore) or on its soil.

As everyone gathered in the sky bar for beer and wine when the ship sailed from port, McCallum was working on an alternative plan. The two best options to avoid Indonesian territorial waters or soil were Dili, East Timor, and Darwin, Australia. Both had the advantage of being on the way to Guam for the Pacific Ocean dives, and the distinct disadvantage of having airports without decent international connections to get everyone who was disembarking home. And so when the *Pressure Drop* sailed from Bali in the late afternoon of April 1, no one was quite sure where it would port next. "No April Fool's joke," McCallum said.

∽

Since the Southern Ocean dives, there had been very good news on the sponsorship front—at least with the one remaining company that was interested. Vescovo had flown to Switzerland to meet with Omega about the luxury watchmaker potentially becoming a sponsor of the Five Deeps expedition.

Vescovo had initially approached Rolex, but they had shown no interest. Likely, the reason was because Rolex had been a sponsor of James Cameron's *Deepsea Challenger* expedition and produced an experimental watch that was a supersized version of its Deepsea watch that Cameron wore on his dive. A marketing tool, the prototype watch was designed to withstand pressure at full ocean depth and was attached to the outside of the submersible. Vescovo reasoned that Rolex probably thought if Cameron, the underwater wunderkind who directed the movie

The Abyss, could only manage a single dive and have his sub never dive again after the experience, who was this Texas cowboy who thought he could dive all five oceans in less than a year with a submarine that was built in much less time? Not wanting to be associated with unrealistic ambition and failure, he assessed that they—quite logically—passed.

It appeared that the joint Rolex–National Geographic effort to back Cameron's submarine and dive was also not successful financially, given that only a single dive had been made. The Swiss watchmaker had continued its relationship with Cameron as part of "Rolex in Cinema." Then, two years after the dive, on the eve of Cameron's *Deepsea Challenge* 3D movie airing, Rolex released a diving watch named the Deepsea D-Blue Dial Diver.

When Rolex gave Vescovo its polite decline, he went to Omega. A chief competitor in the diving market, Omega had actually produced the first commercial diving watch in 1932. Even before the company showed any interest, Vescovo bought a titanium Omega Planet Ocean watch at his local, high-end shopping mall and began wearing it on the dives. McCallum had also been in touch with Omega, where he had contacts, and eventually a meeting was set up.

On March 14, Vescovo flew to Switzerland to meet with Omega's CEO Raynald Aeschlimann to discuss a collaboration. Omega was in the process of commemorating the fiftieth anniversary of the Moon landing in July with its brand ambassador Buzz Aldrin. Doing something with the Five Deeps Expedition was immediately attractive, particularly since Rolex had sponsored Cameron. This was an opportunity to one-up the competition and burnish their marine watch credentials. They also discussed Vescovo taking several Omega watches down in the sub on the dives which Omega could sell, auction, or give away in promotions.

"He got it right away," Vescovo said of Aeschlimann. "He personally apologized for not engaging sooner now that he understood what we were doing." It also helped that the two men also seemed to have an immediate, natural affinity for each other. The cultural fit between the technically obsessed, methodical Omega matched Vescovo's personality and the culture of the Five Deeps Expedition.

Aeschlimann called in his head of project management, Gregory Kissling, and tasked him with building a prototype, full ocean depth watch that could be attached to the *Limiting Factor* on its Challenger Deep dive. After absorbing significant, initial shock because of how little time he had to pull off the feat, Kissling went to work on the design immediately, as the watches would need to be built, tested, and delivered to Guam in just six weeks.

Instead of offering to help sponsor the expedition, Omega offered Vescovo a contract as a brand ambassador, and the Omega logo was put on the *Limiting Factor* front and center as the details were ironed out.

Still, given the permit woes and accompanying uncertainties, the mood at the evening briefing leaving Bali was businesslike. Vescovo said that the deepest point of the Java Trench had been identified and that the plan was to do the solo dive the following day in case of a problem with the Indonesian government.

Lahey passed around a bag of Balinese peanuts as he detailed the changes in the launch and recovery system that had been implemented in the Southern Ocean and improved based on inconsistencies there. The gangplank system using a large rubber bumper had been upgraded. Now, two pontoon-size fenders were set in place just off the aft of the ship to prevent the sub from coming up underneath the bumper and impacting the ship as it had done in the Southern Ocean. Using two bumpers would stabilize the sub in what was called the "dead zone," the area between the ship's two propellers. When the sub was in the water in that space, there was no "wash" to pull the sub away from the ship.

At an earlier meeting, as the team was discussing how to better position the sub away from the stern of the ship on the launch, Vescovo made a head-smacking, "why hadn't we thought of this before" suggestion: "You know . . . on launch, I am in there just sitting and watching everything. But the sub's thrusters have over 200 kg of thrust." He paused, before continuing, thinking through risks and benefits. "When the sub goes fully in the water, why don't I just go max thrust to starboard to pull the sub immediately and quickly away from the ship?" There was a pause, everyone one looked at each other, and it was decided that hell yes, that would actually, really help.

The Triton team had also serviced the winch controls and calibrated the tensions to prevent the sub from twisting as it was lowered into the water. It had added more low-water davits on the hull to adjust the angle of the sub so that it would not be tilted toward the ship. A quick-catch release had been added to the tow lines for faster release. Deck personnel would now wear headsets with voice-activated microphones under their hard hats for more efficient and immediate communication, as opposed to relying on walkie-talkies or what had been the real standard, yelling loudly.

"We are trying to keep the sub from colliding with the stern of the vessel," Lahey said.

"The principle is simple," Vescovo chimed in. "Once the sub is in the water, there is absolutely no reason for it to remain behind the ship. I don't care if I'm 500 meters from the drop point—get it out of there. And we *cannot* launch in a following sea."

Lahey assured him that they wouldn't make that mistake again. "These dives in calmer conditions will set us up for the Challenger Deep, where we will have elevated conditions," he said.

McCallum broke up the seriousness when he asked everyone to return laundry bags, as a shortage had taken hold, and to treat the coffee machine much more gently. It was getting broken frequently. "Don't push the buttons like you are an anxious person pushing a crosswalk button," he said.

~

The first Indian Ocean dive was just the eighteenth for the *Limiting Factor*, a very small number for a first-of-its-kind submersible. Set for noon on the 2nd, the conditions were perfect, with sunshine, calm seas, and little wind. At 9:00 A.M., the lander *Skaff* was deployed to 7,167 meters, close to the sub's target depth in trench depression that was 3 x 1 kilometers wide with a 500-meter "wall" on one side at a very steep 56-degree angle.

Throughout the morning, the Triton team ran through its series of tests on the sub. As usual, there were a few snags—a readout on one battery was not working properly, there was a connection issue with the aft camera, and one downward thruster was refusing to engage. The newly installed manipulator arm was also not functioning properly, but being noncritical that issue was pushed aside for the moment.

As Vescovo was sitting in the control room waiting for the issues to be corrected, he nodded off from jet lag. After twenty years in the military, he had remarked, he could sleep *anywhere.*

Just after 1:00 P.M. Vescovo boarded the sub and was cleared to dive after a seamless launch. "Best yet," Magee said. "I needed that happiness."

Not long after the *Limiting Factor* began its descent, Lahey, in the control room, noted how slow the vertical descent was. The sub was descending at 17 meters per minute, but needed to be going much faster, around 90 meters a minute at this depth. As the water pressure increased and the sub closed in on the bottom, its descent would slow to around 30 meters a minute, but it was far too shallow to be going so slow.

"He's light as shit," Lahey said. "He's not going to make it to the bottom at this rate. He's lost a weight."

Lahey waited one more comms check, but nothing changed. At 1,077 meters, he recommended Vescovo abort the dive, which he did.

With Vescovo back on deck after a near-perfect recovery, Lahey reported that the freeboard weight had ejected uncommanded. Normally, the weight is dropped when the sub reaches the surface to push it above the water line. Losing the 174 kilogram (384 pound) weight made the sub far too light as the water density increased.

Vescovo took it in stride. "I could feel I wasn't going to make it," he said. "I'm getting in tune with the sub. I can feel what it's doing." Then he joked, "Hey, I'd rather have something fail that brings me up rather than pushes me down."

Magee and Blades determined that there had been an electrical fault to the magnet, which caused the weight to release prematurely. The failure came in a non-oil-filled cable that had been crushed as the pressure increased, causing the electricity to the weight's magnet to be cut, which triggered its release as a safety precaution in the event of an actual loss of power. The faulty cable had been spliced because the original wire was not long enough. With multiple dives and compression, the joint had lost adhesion. It was a small example of a failure that had come as a result of Triton rushing to finish the sub in the summer of 2018.

There were other issues on the dive as well. The starboard trunk pump had failed because there was water in the motor, causing an electrical short. "Easy fixes," Magee said.

Maybe so, but Vescovo saw cause for concern going forward.

"We need to continue to think how these repeated deep dives may be stressing connectors and cables to early points of failure," Vescovo told Lahey. "And given the time we have between dives, we should think about wire and connector inspections that need to be done instead of using them until failure. Proactive, not reactive maintenance."

The operational plan was to correct the sub's problems while mapping the western part of the trench one more time, as the middle part had measured slightly deeper on the sonar, in conflict with the existing data available. If the readings were not the same as the first time, a lander would be deployed to confirm the official depth. In fact, it turned out that the center of the trench was deeper than the western part, so that was now the target of the deep dive. The final depth estimates were 7,175 meters for the western point, and 7,188 for the center point, both with an uncertainty of 11 meters. Close, but still deeper.

On April 5, the diving conditions were the best they had ever been. After a launch and recovery on the first dive that worked like "a ballet," as Lahey put it, everyone was confident the goal could be achieved. That morning, everyone drew on Styrofoam cups that would be placed in the trunking and crushed into hardened minis the size of a shot glass by the pressure and taken home as souvenirs of the dive.

As Vescovo prepared to board the sub, Lahey yelled out his standard send-off line: "Have a great dive, Victor."

"Routine," said one of the crew.

McCallum rolled his eyes. "There's nothing routine about sending a man several miles down, something that has only been done four times," he said, adding, "twice by us" and accompanying it with a proud smirk.

Unbeknownst to Vescovo, McCallum had replaced the Marshall Islands flag on the ship's stern with a pirate flag to thumb the expedition's nose at the Indonesian government.

At 9:44 A.M., the *Limiting Factor* began its descent just as the sun tucked behind a cloud. The ship then towed the sub 125 meters to the dive site. At 10:00 A.M., it disappeared from sight.

Ten minutes later, Vescovo encountered a communications issue. Blades instructed him to cycle the power on the starboard modem. This action brought everything back online.

"Holy shit Tom, I'm glad you know what you're doing because nobody else does," Lahey quipped.

The mood in the control room was relaxed. The quarter-hour comms checks were loud and clear, and Vescovo wasn't reporting any issues. Magee was online shopping for a new Ski-Doo. Lahey was cracking somewhat off-color jokes, which everyone had grown accustomed to.

On the dive, Vescovo was descending at 53 meters a minute, fast enough to reach the bottom. On the 11:00 A.M. comms check, he was at 3,475 meters. Blades estimated he would reach bottom at 12:30 P.M. Tapping away on his laptop, Struwe, the DNV-GL certifier, said that it would be closer to 12:45 P.M.

"Jonathan is using a program that takes everything into account, I'm guessing even what Victor had for breakfast," Lahey said, referring to Struwe's quiet but charmingly didactic nature.

Tim Macdonald, who had come on as the new swimmer, took in the atmosphere. "I don't want to sound like a total dweeb, but this is fun," he said.

In the sub, Vescovo had relaxed slightly after crossing through 1,000 meters. On most of the aborted dives to date, the problems had occurred at shallower depths. His only real issue was a minor one, the texting was going in and out. He ate his now-customary tuna sandwich and continually checked his monitors in the "most exclusive café in the world," as he came to call it. Then, between comms checks, he played Bruce Springsteen's "I'm Goin' Down" and Adele's "Rolling in the Deep."

On the way down, the only difference that he noticed between the Indian Ocean and the Atlantic and Southern Oceans was the density of plankton he saw through the viewports. In the Java Trench, it looked like a massive snowfall all the way to the bottom.

At 12:47 P.M., Vescovo reported that he was on the bottom of the Indian Ocean. The official depth would be recorded at 7,192 meters, or 23,596 feet. Cheers went up in the control room.

Lahey informed him that he was 463 meters from the lander *Skaff* and gave him the directional heading.

"Heading there," Vescovo said. "Interrogating the lander."

Vescovo had not successfully navigated to a lander on a solo dive in the deep ocean. Because texting with the control room was not working, he wasn't exactly sure of the coordinates of the lander in relation to where he was. He had somewhat of an idea and knew that it was 650 meters away. Once he started using the thrusters, the control room wouldn't be able to provide him with an accurate heading because of interference. So, he headed in the general direction and continued to interrogate the lander, receiving return pings telling him when he was getting closer and when he was drifting further away. It was like an underwater game of Marco Polo.

After about an hour, the lander's location registered on his sonar. It was about 150 meters away. He kept moving in that direction, skimming along just above the seafloor, peering out the viewports. Along the way, he spotted a plastic bag, not surprising given the massive pollution in the Indian Ocean. When he was about 40 meters away, he saw a blob of bright light. He cycled his own lights so the camera on the lander could get a clean shot of the *Limiting Factor*.

At 2:28 P.M., Vescovo reported to the control room that he had a visual on the lander. Five minutes later, he checked in again but received no response.

The control room could hear him, but he couldn't hear them. Lahey was annoyed. "I don't know why he can't hear me," Lahey said to Blades. "Is this piece of fucking shit not working?"

Blades made a few adjustments. Then, just as Lahey began to fear a repeat of what had happened in the Southern Ocean, Vescovo informed the ship that he could hear them loud and clear.

Vescovo then drove to the lander and stopped just short, working on holding his position against the current for when the manipulator arm worked.

He then decided to undertake some navigational practice. He was, after all, the first person to drive these "roads." He drove 100 meters away, until he was in the pitch black again. He then looked at the sonar to locate the lander, and turned around and drove back to it.

For his last stop, he headed toward the 500-meter wall that the ship's sonar had identified. As he began heading in the direction of the wall, he checked the battery readouts to see how much power he had left. The gauge displayed 5 percent. He had been on the bottom for 2.5 hours, long but not enough to drain the batteries to this degree, so this didn't seem right. He then checked the voltage, which was steady at 140 volts, right where it should be. To determine if he had a faulty readout from the batteries, he blasted the thrusters. When the voltage didn't drop, he knew that the batteries had more than 5 percent capacity, though exactly how much he couldn't be sure.

In the control room, Lahey was getting anxious. On a comms check, he asked Vescovo what his intentions were, which Vescovo took to mean that they wanted him to return to the surface. Adding this to the faulty battery readout, Vescovo decided to call the dive.

With permission to surface granted, he was on the way up.

About two and a half hours later, the *Limiting Factor* surfaced at 5:50 P.M., about 100 meters from the ship. On the horizon, the sun setting behind the distant clouds was so radiantly orange that its reflection off the clouds colored the water pink, a magic hour moment that the TV crew couldn't have scripted better.

The *Xeno* raced over, hooked up the sub and had it in place behind the ship in less than thirty minutes, as fast as a recovery had gone to date. After the swimmer opened the hatch, the first thing to emerge from the hatch was Vescovo's outstretched hand, holding up three fingers: three "Deeps" down.

Later that night, Vescovo talked how it felt being at such a depth for so long. He said that time compresses at the bottom and stretches out on the ascent, saying that going down a minute seemed like three, on the bottom a minute seemed like ten seconds, and on the way up a minute seemed like ten. Having now done three deep dives, what he liked best about them was just being alone in the deep ocean piloting his sub, like, he honestly remarked, any twelve-year-old kid would.

"The solitude reminds me of when I fly my helicopter visually and I'm not talking to air traffic control," he said. "I'm up there looking around not being bothered by anyone. I'm an introvert, way on the introvert scale, so yeah, I like that feeling. Just me and my machine, going somewhere, exploring."

~

The following day was spent preparing for a "reconnaissance and test dive" with Lahey and Jamieson. On the sonar, Bongiovanni had found a large sea wall coming out of a fault line where chunks of the seafloor had been jarred loose. This had been created by massive pressure and breakage of the seafloor. Jamieson had to see it.

To everyone's disappointment, the Triton crew couldn't make the manipulator arm operational. It needed a part that would have to be flown to Guam for the Challenger Deep dive. The arm was becoming something of a bane of existence for the Triton crew.

Jamieson had spent twelve years studying the Hadal Zone, but this was the first time that he would actually go there. "I'm less nervous about doing the dive than phoning my wife and telling her I am doing it after telling her that I would never get in the sub," he said.

On April 7, Lahey and Jamieson dived to 7,180 meters making Jamieson, a Scot, the first British person to reach the Hadal Zone, as well as the deepest diving Brit in history.

During the dive, Lahey drove the sub to the edge of the sea wall feature. Jamieson spotted a large amount of chemosynthetic bacteria mats in orange, yellow, white, and blue, signs of life not fully expected in these waters at such depths, with amazing colors in the otherwise pitch black water. The splashes of color came not from sunlight but from the bacterial mats taking in energy from seafloor itself. Lahey deftly maneuvered the sub under an overhang on the rock wall. Jamieson called it the most unbelievable geological feature he had ever seen. Sometimes the cold cynic, the scientist was completely awed by what he saw on the dive.

"The geological complexity was amazing," Jamieson said. "I even discovered a new species of snail fish with my own eyes, which was something I will never forget. At one point we were homing in on what appeared to be chemosynthetic sponges hanging in a group, which itself was amazing, even more so when we realized we were under an overhang. The dive was just magical."

The scientific finds not only revitalized Jamieson's spirits, they were the most varied of the expedition to date, despite having only two successful dives and seven lander deployments in the Java Trench. The lander cameras captured six species of marine life. These included a jellynose deeper than 6,000 meters for the first time; a brotulid (or cusk eel), which might have been the fabled abyssobrotula, a new species of snailfish; the deepest larvaceans ever filmed, and a giant amphipod (*Alicella gigantean*), seen for the first time in the Indian Ocean.

The cameras also filmed the deepest octopus ever—a dumbo octopus (*Grimpoteuthis*)—beating the current records by 1,802 meters. "At least now we know cephalopods [octopus] are indeed Hadal, and the depth extension from the Java Trench increases the entire cephalopod groups benthic habitat from 75 percent to 99 percent globally," Jamieson said.

In a new twist, one group of previously unseen amphipods proved to be predators. They were captured by a lander camera eating a mackerel in one bite, leaving behind a small pile of bones. Snailfish, which are transparent and identified by two mouths, were known to be abundant in the deepest parts of the Hadal Zone, but they had never been filmed in action until now.

Most exciting was the money shot from one of the landers: a marine creature that Jamieson nicknamed "Snoop Dog." It looked like a gelatinous blue balloon with a tail-like long string attached and bore a striking resemblance to a dog's head. Jamieson was uncertain at first what it was. Upon further examination, he identified it as a new species of jellyfish, a tunicate sea squirt (*Ascidian*).

"The mapping of the Java Trench was also incredible in terms of geological resolution," he said. "I kept taking photos of the multibeam screen of these large, four-kilometer-wide, heart-shaped depressions on the overriding plate and sending them to Heather (Stewart, the geologist who did not make the trip), asking if they were tsunamigenic landslides, which they were."

The debate over the scope of the science that the expedition could achieve was coming into focus. At sea during the trip, Vescovo had a conference call with the chief of content from TED Talks, who had invited him to be interviewed on stage about the expedition at their next conference in August in Vancouver. The TED executive explained that they would put up pictures on a screen of the "big discoveries" and the "cool creatures" that Vescovo had found while piloting the sub in each ocean.

Vescovo explained that the Five Deeps Expedition was foremost a *technical* mission focused on building the first full ocean depth submersible and then successfully operating it on multiple dives to the bottom of the five oceans. No one had ever done that, he emphasized.

After asking Vescovo several times what he had found with the sub, the TED executive understood what Vescovo was saying. "So when you go down, it sounds like there's nothing there but a lot of mud," he said.

Vescovo replied, "Well, not exactly, we have made finds, but that's the nature of the trenches."

The two agreed that he would talk about the technical aspects of the mission and bring images of any new sea creatures that had been filmed.

"He was making a point from a popular science standpoint, that it's dark and the sub hasn't really found anything," Vescovo said after the call ended. "This is the difference between popular science and hard science. Many things we can find you can't see, like new genetics sequencing and understanding the evolutionary drift at lower levels where most of the ocean's life is. We have climate models that aren't predicting what they said would happen because we don't fundamentally understand how the ocean interacts with the atmosphere, because we don't have data. We've been to just 20 percent of the ocean, because 90 percent of it is below 1,000 meters. The sub is a technological achievement to open the door where people can begin to use it where we couldn't go before. It's just not that flashy. But it can move the needle, and the next owner of the system can pick up the ball, go to *any* point on the seafloor and do ten, twenty, thirty dives at each site if they want. That is a capability that has never existed before."

Science, of course, is not achieved in one go. It requires repeated study and observation, but it must have that first breakthrough to take further steps. This became what the expedition was trying to achieve in its dives. As a Chinese proverb says, "A journey of 1,000 miles begins with a single step."

❧

During the voyage, Vescovo had been in touch with his college roommate at Stanford and close friend, Enrique Alvarez, a cyber-crimes investigator with the FBI. Alvarez was in contact with a U.S. Navy attaché in the Indonesian embassy to make them aware of the situation with the permit in case the Indonesian government tried to interfere with the expedition in any way. The naval attaché emailed Vescovo that no contact had been made by the Indonesian government and everything appeared fine.

To play it safe, the expedition decided against returning to Bali. Instead, the *Pressure Drop* would sail to Dili, East Timor, and everyone leaving the ship would fly back to Bali to make their original flights home. They doubted the Indonesians would detain them while individually just making international air transfers in Bali. Docking the ship in an Indonesian port, however, might have been too tempting for the government authorities. They might demand a search of the ship for marine samples or just generally just give the team a hard time. It wasn't worth the risk.

The ship spent three days sailing just outside Indonesia's territorial waters and fairly close to land to steer clear of a cyclone heading into the Indian Ocean off the Australian coast. The balmy air and billowing clouds of the day turned into an orange kaleidoscope as the sun tucked in on the horizon each evening. Gazing across the endless ocean to a slight curve in the distance, it felt like you could actually see the Earth rotating.

The *Pressure Drop* arrived in Dili on April 10 at 6:30 A.M. and anchored in port. No less than fifteen customs officers boarded the ship and met with Buckle in the galley. Buckle brought the box of passports, all carefully examined by multiple officials. Once everyone was cleared, they boarded a skiff to shore.

From the ship, the small island country with rolling green hills resembled a vacation destination. But once on land, it was an underdeveloped, chaotic mess of cars coughing out smoke and signs of poverty everywhere. The port agents suspiciously but politely loaded the suitcases into several cars and flatbed trucks in which passengers sat in the back clinging to their luggage. The team members were driven to a ramshackle airport for the flight to Bali.

After clearing airport security, Jamieson and two of the crew found a bar with a Formica counter in the back corner of the rundown terminal. Despite the early hour, they had a couple rounds, toasting to the expedition's success on its "freedom of navigation" mission in the Indian Ocean.

"The dive was a career highlight," Jamieson said. "Funny how in just a week or two I have gone from wanting to quit to thinking I have the best job in the world."

∞

Vescovo held off on having a press release issued until Buckle advised him that the *Pressure Drop* had completely left Indonesian waters—including its EEZ, which occurred on April 15. Buckle had been adjusting the ship's tracker periodically, resulting in the location on public tracking sites being a day or two behind the ship. "I imagine the tracker page on the website does look like I've been hitting the rum," Buckle emailed Vescovo.

With the Indonesian presidential election looming on April 17, Vescovo's PR firm, the Richards Group, issued the press release detailing the Indian Ocean dives on the 16th to stay under the news radar in Indonesia. Regarding science finds, the release stated that the expedition had made only general films of the dives to test the camera systems before their next, extreme, descents in the Mariana Trench. Technically, this was true.

That same day Jamieson finally received a formal letter from the Indonesian Foreign Research Permit Secretariat informing him the research application had been "categorized as unrecommended objects and activities for foreign researchers," and therefore denied. No further explanation was given.

A few days later, Jamieson received another letter dated earlier from the same organization signed by the same person that stated the research application was not approved "due to national security consideration." Again, no explanation was given as to why the scientific research of the expedition could somehow prejudice national security. Vescovo commented to McCallum by email that this was actually an absolutely invalid reason to disapprove the application since marine operations in an EEZ were not allowed to be denied based on "national security"—that was clearly not an allowed reason for denial according to the UN Convention on the Law of the Sea (UNCLOS). They could only do that, he claimed based on his navy experience, within their twelve-mile territorial waters. The Indonesians, he commented—his disgust apparent even in an email—apparently didn't even know the provisions of the Law of the Sea they had signed onto.

Months later, Newcastle University officials contacted Jamieson and said that it had received a threatening letter from the Indonesian government because of his affiliation with the university. Jamieson thought his position might be in jeopardy if things escalated. Vescovo hired Mike Lax, a London-based UNCLOS attorney, to respond on behalf of Jamieson and the expedition.

Lax wrote a detailed legal defense of why the expedition should have been granted a permit under the UNCLOS guidelines. He also pointed out that Indonesia is a signatory to the International Hydrographic Organization's charter that requires all members facilitate research in their waters for the benefit of science. The Indonesians had plainly violated this obligation, while garnering whatever benefits they obtained from the organization's efforts. He asked that the government reconsider the application retrospectively, and repeated the offer to share the limited results of the expedition's findings.

The Indonesian government never responded to Lax. Instead, three months later the Indonesian government passed a law that "imposes criminal charges on foreign researchers who steal biodiversity samples with a prison term of two years and a fine of up to 2 billion rupees," roughly $143,000.

Upon hearing the news, Vescovo remarked to Jamieson: "Good luck to them ever getting anyone to ever do research with them in the future. I, for one, will never go back. To hell with them."

CHAPTER 15

DARK, DEEP, AND DANGEROUS

The notable British science fiction writer Arthur C. Clarke once said that it was inappropriate to call our planet "Earth" when more than 70 percent of it is ocean. Of the oceans, the Pacific Ocean dominates, covering more of the planet than Earth's entire land mass, some 64 million square miles in all. Its deepest point is deeper than those in any of the other four oceans. This area, lying in the Mariana Trench about 200 miles southwest of Guam, is named the Challenger Deep after the British ship that first recorded its depth in the 1870s.

The Challenger Deep is easily the most challenging place on the globe to reach. Only three humans on two dives had been there, U.S. Navy Lieutenant Don Walsh and Jacques Piccard in the *Trieste* on January 23, 1960, and James Cameron on a solo dive in the *Deepsea Challenger* on March 26, 2012. In April 2019, Victor Vescovo would attempt to become the fourth to visit it and second to dive it solo as one of a series of planned dives.

McCallum had arranged for Walsh to be on board for the Five Deeps Expedition's mission to their fourth deep. At eighty-eight, the spry, witty Walsh still traveled the world participating in oceanographic events. He had been a guest on the *Mermaid Sapphire* when Cameron completed his dive to the bottom of the Challenger Deep.

Vescovo met Walsh for the first time in Guam the day the *Pressure Drop* set sail. As a former navy reserve officer, Vescovo had immense respect for Walsh—and vice versa, as Walsh was duly impressed by the expedition that was bringing deep diving back into focus. As a token, Walsh presented Vescovo a Navy "Deep Submersible" qualification badge. "I'm not going to put it on until I earn it," Vescovo said.

Leading up to the Challenger Deep dives, there were yet again logistical snags, but nothing that appeared to be a real impediment. McCallum grappled with the slow-moving bureaucracy up to the last minute to secure the necessary science permits for the Mariana Trench. He was still waiting for clearance for a planned dive to the bottom of the neighboring Sirena Deep, where no human had ever been and the third-deepest spot on the planet (the Horizon Deep in the Tonga Trench would turn out to be second).

There was also an issue with the *Pressure Drop* porting in Guam and returning after the dives. The guidelines for porting that McCallum initially received did not include two pages stipulating that the expedition hire a local port agent to draft a Vessel Response Plan (VRP) for accidental oil discharge. When McCallum discovered this, he contracted with a local agent at a cost of $30,000 to have a VRP drawn up. The VRP would then need to be approved by the U.S. Coast Guard in Washington, D.C., and sent back to the Coast Guard office in Guam. While the paperwork was being processed, the ship was granted a onetime exemption to port.

To try and ensure that there was no issue returning to Guam, McCallum and Walsh hosted forty U.S. Coast Guard team members on the ship. They went through the vessel from stem to stern, in awe of the expedition. McCallum also collected several of their military unit "coins" and badges for Vescovo to take to the bottom in the *Limiting Factor.*

An added dimension to the mission was that the Omega team, led by Gregory Kissling, had designed and manufactured three full ocean depth watches in just over six weeks. The casings were made from titanium taken from the cut-outs for the viewports in the hull of the *Limiting Factor.* The three prototype watches were named Omega Seamaster Planet Ocean Ultra Deep Professional, serialized as FOD (Full Ocean

Depth) X1, X2, and X3, and were just a bit thinner than the deep ocean watch Rolex had made for Cameron's mission. Triton's Hector Salvador had spent Easter Sunday testing the watches to DNV-GL standards of 15,000 meters in Triton's pressure chamber in Barcelona.

"This project shows how Omega (with the great support of the sister companies of the Swatch Group) and its powerful DNA filled with pioneering spirit can master such challenges," Kissling emailed Vescovo four days before the *Pressure Drop* was to set sail for the Challenger Deep. "It was an audacious thing for us knowing we started the project only mid-March 2019. We dared to supply a new conception that has never been seen in the watchmaking industry and this was the winning one. And of course, fingers crossed for the Challenger dive!"

Kissling himself flew to Guam with the watches—or as the Omega team insisted on calling them: "timepieces." The plan was for two watches to be attached to the sub's manipulator arm, which Vescovo would photograph and film from the inside of the sub with his Samsung phone camera. The third watch would be attached to the lander named *Skaff*.

"We felt confident the timepieces would withstand the pressure because we had tested to 25 percent beyond full ocean depth to certify them with DNV-GL to full ocean depth," Kissling said. "But with something that has never been done before, you can't help but be anxious."

As part of his ambassadorship arrangement, Omega was giving Vescovo one of the FOD watches, as well as fifty titanium Seamaster Diver 600 meter watches for the expedition team. Vescovo planned to take an additional one hundred and fifty Seamaster watches to the bottom of the Challenger Deep in the *Limiting Factor* for future Omega promotional use. Omega was having special backs with the Five Deeps Expedition logo made to commemorate the dive for the crew's timepieces. The only proviso was that the recipients had to agree not to sell them. "As if!" threw out Triton's Frank Lombardo. "I'm passing this on to my son, and then to his son."

On April 26, just before 5:00 P.M., the *Pressure Drop* sailed from Guam's Apra Harbor with a complement of forty-eight on board. Dr. Glenn Singleman had rejoined the expedition, a precautionary measure because of the elderly Walsh's presence. On the science front, in addition to Jamieson, Dr. Patricia Fryer had joined the expedition for the dive. Fryer, a researcher and professor at the University of Hawaii, had spent thirty years studying the Mariana Trench. She had actually discovered and named its second deepest point, the Sirena Deep—or rather let her students name it.

The excitement level at the departure briefing was high. There was a sense that the expedition, which had struggled on so many fronts for so many months, was on the cusp of achieving something special in the history of deep ocean diving.

In the departure briefing, McCallum laid out how operations would be running around the clock. Mapping would occur in the overnight hours. Because of the length of time needed to reach the bottom—four hours—lander deployments would begin at 3:30 A.M. The team needed data from the landers on the bottom so they could precisely determine how many weights to place on the submersible. The sub would then descend to meet the landers already on the bottom. "We are in the major leagues now," he said.

Vescovo added: "We are assaulting the Challenger Deep with a record level of technology. Sonar, sub, all three landers . . . it's unprecedented."

Lahey, however, was slightly less high-strung than usual. "We have all the poker chips in the center of the table," he said. "This is as good as it gets in deep sea diving."

Buckle, however, was concerned that complacency could set in after the successful dives in the Indian Ocean, because the expedition had slotted twelve days for the five Mariana Trench dives. "After the meeting where Rob talked about how intense it was going to be, Patrick huddled with the Triton crew and said, 'We've got lots of time. We can take it slow. We have days built in for maintenance and extra dives,'" Buckle said later that evening. "The way Triton works, that's not the approach we should be taking."

The weather forecast was extremely favorable. Nevertheless, the plan called for Vescovo to attempt the solo dive when the ship reached the Challenger Deep, in the event of failures with the sub or sudden changes in the weather. That dive, after all, was also the primary goal of the mission.

⁂

In the days leading up to the Challenger Deep leg, Vescovo had been emailing back and forth with James Cameron. The filmmaker had followed the development of the *Limiting Factor* through Lahey, a longtime friend, and asked for Vescovo's email. Indeed, Cameron alumni were well represented on the ship.

Buckle had served as captain of Cameron's Deepsea Challenge expedition. McCallum was its expedition leader for the test dives. Fryer, the geologist, was a member of Cameron's science team. Singleman and P. H. Nargeolet had worked with Cameron on other expeditions. Even Marcos Benavides on the kitchen staff had worked the galley on Cameron's ship.

In his first lengthy email, Cameron wrote Vescovo: "You and I are the only two people who have operated programs to dive humans to Hadal depths in almost 5 decades, so as one deep sub pilot to another, I just want to say I know what it takes to go that deep, and I know what it feels like to be there, alone with all those miles of water over your head."

He provided many details of his dive, including his final depth of 10,908 meters (35,787 feet) plus or minus 5 meters, which was a post-expedition recalculation based on the UNESCO formula and his own CTD (conductivity, temperature, and depth) data. "So that's the depth to beat," he wrote.

Cameron also opened the door to discuss how Vescovo and the *Limiting Factor* could participate in OceanX, Ray Dalio's expedition being filmed for a National Geographic Channel show, on which Cameron was a partner. He suggested the possibility of Dalio taking a lease-to-buy approach and diving the *Limiting Factor* from his ship, the *Alucia*.

The two arranged a call from the ship on the day before Vescovo's solo attempt. Vescovo phoned Cameron, who was on a sound stage working on his *Avatar* sequels. When Vescovo introduced each of the Cameron alumni gathered in the room, along with Walsh, Cameron quipped: "With one grenade you could take out most of deep ocean exploration."

During the twenty-minute call, Cameron was encouraging and gave Vescovo advice on where to go. He said that he hoped that Vescovo could bring back a rock for Fryer to examine, something he had been unable to do. He spoke in detail about his route on the bottom, talking about how flat the terrain was, and said that after spending three hours on the bottom, he had been forced to abort the dive because of thruster issues. He had thought he would be able to dive again the following day, but there turned out to be too many issues with the sub, along with bad weather closing in and other commitments he had.

"The only piece of advice I would give, even if you have to write it in your dive plan, which I did, is to stop and look and think," Cameron said. "You tend to get very hermit like, running the checklist, I have this to do and that to do. Give yourself a moment, stop and look out the window and think about what it means—and then get on with your checklist. That's the moment you will remember for the rest of your life."

"I appreciate the support you've given about doing the dive and the psychological support," Vescovo said. "It's great sitting next to the first person who went down to the Challenger Deep and talking to the [third] person, and hopefully we can continue passing the baton with technology and everything that we are learning."

"You've taken the ball further down the field in terms of technology and repeatability, and it's really great what you are doing," Cameron said. "I'm glad you are going to swivel this world spotlight back on this instead of all the bullshit politics that everybody is so absorbed in."

"There's a lot of attention on space, but I think we need to explore near space," Vescovo said.

"Well, good luck," Cameron added. "And as we always said on the dives, 'May your surfacings always equal your descents.'"

⁂

The Challenger Deep has three "pools" broken up by ridges. Walsh dived in the Western Pool, while Cameron dived in the Eastern Pool, as would Vescovo. Mapping lead Cassie Bongiovanni and her teammate Seeboruth Sattiabaruth had extensively mapped 13,000 square kilometers of the trench. Their initial findings determined the deepest point in the Western Pool was 10,938 meters, while the deepest point in the Eastern Pool was 10,943 meters. Both carried a possible twenty-meter uncertainty and were expected to change with the application of a full-ocean sound velocity profile from the sub and the landers. This data would be applied to the mapping data to arrive at an accurate depth.

In the early morning hours of April 28, all three landers were deployed. The *Limiting Factor* dive was set for 8:00 A.M. With all systems a go, Vescovo boarded the sub with a duffle bag of Omega Seamaster Planet Ocean watches in a container on the seat next to him, carrying a retail value of over $1.1 million. Two of the specially made Ultra Deep Professional watches, costing into six figures to manufacture, were mounted on the outside of the sub. "With all those watches, he had better not miss a comms check," Lahey deadpanned.

After a focused LARS crew conducted what was now dubbed a "ballet launch," the sub began its descent. The internal cameras recorded Vescovo saying, "It all comes down to this after four years of planning, testing, and diving . . . this is the big one."

His dive was going smoothly until 11:15 A.M. With the *Limiting Factor* crossing through 8,700 meters—more than two-thirds of the way down—Vescovo noticed that the range to lander *Skaff* was decreasing far more rapidly than was normal. After a few more communication "pings" to the lander, Vescovo realized with frustration the obvious development: *Skaff* had unexpectedly released from the bottom. The lander was headed up far too early. This created a potentially dangerous situation—the possibility of an actual "mid-ocean" collision. The *Limiting Factor* had been dropped in the same place, meaning that the ascending lander could be heading up in the same water column as the descending sub.

Almost at the exact same time Vescovo realized what had happened, Lahey radioed Vescovo. "Be advised lander *Skaff* has left bottom," he said.

Vescovo acknowledged, but there wasn't much anyone could do about it. "I always worry about midair collisions as a pilot but not in the Mariana Trench," he said to himself. "Strange things happen down at this depth." Still, he looked down through the bottom portal for the telltale bright light of the lander, his hand on the thruster controls ready to make a quick move. Vescovo was incredulous that he might have to do a last-minute evasive maneuver to avoid colliding with his own lander. Still, he thought, things moved slow enough that it shouldn't be too much of a problem if it did happen.

The control room sat on edge and waited, as the large monitor that charted the locations of the lander and the sub showed the depths of the *Limiting Factor* and *Skaff* closing in on one another.

"The comms room looks like Rembrandt's *Night Watch* painting," Walsh said, referring to the master artwork with concerned military officers standing around.

A tense twenty minutes ended when the depth readings indicating that the lander had passed the *Limiting Factor* without incident, which Vescovo also saw on his monitor.

Ninety minutes later, Vescovo was approaching the bottom. When he was 20 meters away, he saw a soft glow coming off the seafloor. He slowed his descent, as the moment was at hand. Then, at 12:38 P.M., a small cloud of brown sediment kicked up around the submersible. The *Limiting Factor* had done it.

"Touchdown," Vescovo said to himself.

As the dust settled, Vescovo peered out the viewport. He paused for about half a minute and took a deep, relieved breath. He finally radioed the control room: "Surface, *LF*, at bottom. Repeat at bottom. Thank you all, you made this happen."

The place erupted. "Congratulations Victor," Lahey radioed.

"Congratulations to you all . . . beginning exploration of the bottom" came his reply.

Vescovo spent just over four hours on the bottom, the longest by well over an hour and a half. He was surprised by how much life he saw with his own eyes through the viewports. He spotted a sea cucumber and a handful of amphipods, as well as a few odd-looking sea creatures that he would need Jamieson to see on film to identify.

At one point, he also zeroed in on a piece of what appeared to be a plastic container, or bag, of some kind—almost the size of a small potato chip bag. "Nature doesn't do straight edges," he said, shaking his head. "That's contamination of some kind." He attempted to grab it with the manipulator arm, but it floated just out of reach before getting lost in the kicked-up silt.

After four hours of navigating in all different directions at the bottom of the ocean, the *Limiting Factor* was running low on power, so Vescovo decided to surface. Before he did so, though, he took Cameron's advice and stretched out in the capsule, quietly ate a tuna fish sandwich, and slowly drifted along the bottom in total peace and quiet.

Finished, he jettisoned his drop weight and radioed the surface he was coming home.

His journey up lasted well over three hours, and at 7:42 P.M., he was cleared to surface. The sky had turned dark with only a dim moon that shone through clouds, lighting the water, making the recovery somewhat challenging. The sub was fully lit, but clinging to the top of the sub, Tim Macdonald, the swimmer, struggled to attach the tow line. After three tries, he was successful, and the sub was towed to the stern.

"That was sporty, boys," an overjoyed Lahey said, backslapping his guys.

At 8:20 P.M., some twelve hours after the dive began, Vescovo climbed out of the sub with four fingers extended. "Four! Thanks for coming to get me!" he yelled.

Back on deck, he was presented with a Hawaiian lei by Fryer, and high fives and hugs were doled out to everyone on the team. The first person he shook hands with was Don Walsh.

Struwe, the DNV-GL certifier, asked Vescovo how much power he had left when he surfaced.

"Pretty much none!" Vescovo said.

Blades, the chief electrician, smiled. "That's what an explorer will do, push it to the limit," he said.

Even Walsh marveled at the dive. "He has more balls than a bowling alley," he said.

And so, 21,645 days after Walsh and Piccard's dive, and 2,589 days after Cameron's, Vescovo had become part of an elite club of four men—just the second to solo—who had visited the deepest point on the planet, putting them closer to the center of the Earth than any human had gone. (The actual closest point, according to geologist Heather Stewart, is in the Gakkel Ridge in the Arctic Ocean under the ice cap, where, of course no one has ever been.)

After asking about the magnitude of the dive, the film crew asked him how he felt physically. Standing on the deck gripping the ice axe that he had taken to the top of the world on Mount Everest in 2010 and now to bottom of the ocean, he laughed. "My legs are pretty tight," he said.

∞

On May 1, Vescovo prepared for a second solo dive to the bottom of the Challenger Deep, which would make the *Limiting Factor* the only sub to dive it twice. As he headed across the deck to board the sub, he said, "We're going to try and make history—again."

On the launch, the sub had to be towed to the drop point. It took nine long minutes. By the time it reached the targeted point, its ballast tanks had taken on a decent amount of water. As the sub began to submerge prematurely, Macdonald, the swimmer, fought to unclip the sea anchor and remove the railings.

"Release! Release!" Lahey yelled.

"Tim, you okay!" yelled Triton's Magee, who was running deck operations.

A frantic minute passed as the sub—with Macdonald still on it—disappeared from the surface and the thought that he had become entangled in the cables and railings flashed through everyone's mind. Seconds later, with the sub fully underwater and out of sight, Macdonald popped up to the surface.

Back on the ship, Macdonald, whom Vescovo always admitted probably had the most dangerous job on the crew, shook it off. "I wasn't going to quit because they would have aborted the dive," he said.

The film crew was thrilled. The footage from Macdonald's helmet camera as he was pulled under with the sub was some of the most dramatic yet.

Vescovo's descent was uneventful. One thruster on the port side had suffered water ingress on his first dive and was not functioning, which he knew when he left the surface. Early in the dive, the port side headset went out, so he switched to the starboard. Vescovo reported it, recommending that the dive continue, and Lahey agreed. Neither issue was an abort condition.

At 11:58 A.M., he touched down on the bottom close to the lander *Skaff*. It was the first repeat dive—ever—of the Challenger Deep by a vehicle or a person.

For something so remarkable, there was a feeling that the team was now so dialed in that they could dive the Challenger Deep over and over again. This was something. The team of Walsh and Piccard together and Cameron alone, after all, had only done it once, and for various and differing reasons, never attempted it again.

When someone brought up how well the dives had gone, Lahey, perhaps the proudest person in the room, became philosophical. "You have to be absolutely humble with the ocean," he said. "You need to take any endeavor in the ocean with deep respect. I take nothing for granted and we can't either, especially at this depth."

Vescovo stayed down for three and half hours. He spotted what looked like a piece of underwater fiber optic cable and took a picture of it with his phone. He worked the manipulator arm and late in the dive grabbed a rock. However, he didn't have enough battery power to drive to the lander and place the rock in the basket so he had to release it. But at least the manipulator arm was working.

At 3:30 P.M., he dropped his surfacing weight and headed for the surface. On the way up, at around the 10,000-meter mark, an alarm went off indicating a battery fault. Vescovo thought he might have even smelled a whiff of burning insulation. Moments later, the main lights in the sub went out.

Vescovo's heart rate spiked. He immediately realized that the emergency power had taken over, and with that, many nonessential services like cameras and internal lights had gone off-line. He had used up all of his regular internal power sources but he wasn't overly concerned, though, because he knew that the *Limiting Factor* actually had *twice* as much emergency power as regular internal power. That had been a DNV-GL safety requirement to obtain commercial certification. He knew he probably had at least 12 hours of emergency power—far more than enough to get to the surface and on board safely. In fact, the craft didn't need *any* power to safely ascend.

"Surface," he radioed. "*LF* has lost power, is on emergency battery power only."

"Roger, understand, we will track you from the surface," Lahey replied.

Vescovo attempted to reset the internal battery circuit breakers one time each, but they both flipped back off. He opted to leave them alone, lest they create a fault in the emergency power. He could feel that he was ascending at a good clip based on seeing the particles outside the window. However, he no longer had any depth information because the CTD data in the sub was off-line. His comms were clear, so he decided to remain on emergency power until he reached the surface.

With the power out, the heater had stopped. The temperature rapidly dropped from 50 degrees Fahrenheit to around 33. Vescovo pulled on his stocking cap and gloves. It was uncomfortable, though not consequential. He thought, at least this is better than the mountains. But then, at 6,500 meters, an independent Analox unit blasted a high-CO_2 alert scream.

This meant that either the two CO_2 scrubbers in the capsule were not working properly, or that he was receiving a faulty readout. The normal level of CO_2 concentration in the ambient air was 0.04 percent. The alarm was set to trip at 0.5 percent. Vescovo looked at the readout, which was rising quickly and soon reached 0.72 percent. At 2 percent, the capsule's atmosphere would reach a danger zone that could result in sharp headaches, chest pain, respiratory and cardiovascular effects, and visual and other central nervous system effects. It wouldn't result in death, but could make things very uncomfortable.

He radioed the surface that "CO_2 level is spiking." Taking no chances, he pulled out a lithium blanket packet designed to increase CO_2 absorption. He took out a pocket knife from his flight suit and very quickly cut open the package and hung a blanket like a curtain.

Lahey radioed him. "*LF*, what is your CO_2 reading?" he asked anxiously. "Confirm both scrubbers are running. Power up secondary unit."

Vescovo did. He rechecked the scrubbers. The CO_2 climb rapidly stabilized and soon began to drop. "Control, be advised CO_2 level back down to zero point three two," he said.

"Roger that," Lahey said. "Okay man, see you on the surface . . ."

The troubleshooting over, Vescovo muttered, "Well, I think I'll finish my Kit Kat."

Vescovo reasoned that the battery outage had triggered a false reading, as it would have taken more than an hour to spike to such a high level. Nevertheless, being alone in a dark titanium sphere with the weight of the ocean on top of him and an indication that the CO_2 level was climbing rapidly was not a comfortable feeling, even for a highly experienced explorer.

But what he knew was that sometimes the hardest and most vital thing to do is to stay calm. Part of the ability to deal with potentially life-threatening crises goes back to the Exploration Gene Theory. In addition to needing to seek the extra thrill in life, Singleman explained that people with 11 copies of the DRD4 gene on chromosome 11 also react differently when tested. Rather than being overwhelmed and having their fear response take over, they respond with an orientation reflex. Their heart rate actually drops, they switch into a problem-solving mode based on their preparation and planning, and reorient themselves to the new situation—even if it is a new challenge miles underwater.

The remainder of the ascent went smoothly. The *Limiting Factor* surfaced at 7:10 P.M. and was recovered without issue, completing the fourth dive and third solo dive in history to the bottom of the Challenger Deep. But, as often occurred on the expedition, success was followed by a problem. In this case, the lander *Skaff* had failed to surface and lost contact with the ship.

Aside from a $300,000 lander missing, *Skaff* had one of the six-figure, Omega prototype watches attached to it.

Tensions quickly mounted. The landers had been deployed without a third recovery feature—a galvanic release that relies on metals corroding and releasing the weights—to bring them back up in the event the electronic communications and timer failed to do so. The galvanic release set for 24 hours had failed the day before and caused the lander to ascend after just six hours. It was removed at the request of Jamieson and against the wishes of Lahey. Therefore, *Skaff* had no means of releasing from the bottom.

This had been the subject of a heated debate between Jamieson and Lahey the night after Vescovo's first solo dive.

Lahey had been pushing for the landers not to be deployed without the galvanic release working properly, reminding everyone of the lost lander during sea trials in the Bahamas.

Jamieson had insisted that the likely cause of that mishap was a bad timer. All the timers had been replaced with much more reliable ones from a different vendor.

Vescovo had listened to both sides. "Have the timers failed since the Bahamas?" he asked.

"No," Jamieson said.

Magee had added that the new timers were powered by a secondary battery so that if the primary timer failed, the lander would still release.

"If they are preventing the mission from being accomplished, it doesn't matter if we have them," Vescovo reasoned. "Forget the galvanic release. I don't trust it anymore."

Jamieson jumped up from the table. "I'll take that as a final answer."

The following day, all three landers had been deployed and recovered. But now, on the night of Vescovo's second deep dive, *Skaff* was, quite literally, stuck in the mud in what was arguably the most remote, inaccessible place on Earth.

❧

The next dive, set for May 3, was the certification dive required by DNV-GL. Lahey and Struwe would dive the *Limiting Factor* to full ocean depth. The dive was contractually supposed to be 100 meters less than Vescovo's depth, but circumstances had changed. He no longer cared if Lahey and Struwe went to the same depth, he needed them to recover the lander and the Omega watch on it, if at all possible. So now the dive would also be an audacious salvage and rescue mission to free the lander from the bottom—by far the deepest such marine recovery operation that anyone had ever attempted, or *could* ever attempt for that matter.

The implications were apparent. If the *Limiting Factor* could go to the bottom of the deepest ocean and find the lander, it could go into any ocean and find something where the coordinates were known, such as a downed aircraft or sunken ship.

Buckle had the exact coordinates where *Skaff* had been dropped, though there was no way to know just how far the lander had drifted during its 6.5-mile-plus descent. Before the dive, another lander, *Closp*, would be dropped in the same place as *Skaff* so that they could conduct a "square" search pattern. This would give the sub an active lander to navigate to while searching for *Skaff*. The theory was that they would drift similarly and be close to one another.

Lahey and the normally reserved but now visibly excited Struwe left the surface just after 8:00 A.M. On the way down, they interrogated *Skaff*, and the lander seemed to almost plaintively respond with a single "ping" from its certainly dying batteries after two and half days on the bottom. This meant that the two landers were between 100 and 200 meters apart, raising the confidence level in the control room that *Skaff* could be found.

The sub stopped its descent just short of the bottom at 12:48 P.M. and headed for *Closp* to position themselves to find *Skaff*. Forty-five minutes later, they reached *Closp*. Soon after, Lahey reported that they could see *Skaff* on their sonar and were heading that direction. Fifteen minutes later, Lahey checked in again: "At *Skaff*!"

Smiles and laughter abounded in the control room seven miles above.

Now, the sub needed to nudge the 800-pound lander loose from the seafloor. The lander was resting at an angle with part of its base wedged in the sediment. Its single strobe light was miraculously still on. Struwe extended the manipulator arm, which wasn't fully cooperating but extended from the sub. Lahey drove forward until the arm made contact with the lander. He then blasted the sub's thrusters, attempting to dislodge the lander. Struwe was glued to the viewport watching the progress.

After two or three heavy blasts on the thrusters, Struwe said, "He's going up . . . he's on its way up!"

Lahey and Struwe shook hands. "Surface, *LF.* Lander has released," Lahey reported.

Celebration broke out in the crowded control room. "Yes!" Magee shouted.

He and Vescovo double high-fived.

"A technological *tour de force*," Vescovo said. "It's the deepest marine salvage mission in history. And likely for all time."

Blades estimated the lander would be up around 7:00 P.M. Now the question was: did the Omega watch survive almost three days at full ocean depth?

Lahey and Struwe had a mini celebration in the sub on the way up. Lahey played "Gimme Shelter" on his phone, as the two snacked on chef Manfred's apple strudel.

While the sub and the missing lander were on their way up, on what was shaping up as one of the most successful days of the expedition, Vescovo went outside to get some air and take stock of the Challenger Deep dives. Leaning against the railing at the starboard muster station, he contemplated whom he would approve for the fourth and fifth dives. For everyone on the ship, these deep dives were badges of honor that could not be duplicated, once-in-a-lifetime experiences that carried gravitas in the oceanographic world.

The fourth dive was Triton's contractual discretionary dive, the one that Bruce Jones had wanted his wife to go on. A mutual and logical decision had eventually—if painfully, on Jones's part—been made that sub designer John Ramsay would be the passenger, with Lahey piloting. Vescovo was contemplating letting the pilot be Magee, who had run deck operations since the beginning and been so instrumental in keeping the sub up and running.

"I know Patrick doesn't think I trust anyone else but him to pilot the sub, but I do trust Kelvin," Vescovo said.

The fifth dive would be in the Sirena Deep, about 140 miles northeast of the Challenger Deep and about 200 meters shallower, now that the permit had been approved—aided in no small measure by an email from Patty Fryer, discoverer of the deep.

"Patty wants a rock and Dr. J is complaining that there is no significant life in the depths of the Mariana Trench for him to study and so he wants to see some in Sirena," Vescovo said.

After mulling it over, he made the decision to make the fifth dive a science dive with Jamieson in the Sirena Deep.

The *Limiting Factor* surfaced and was back on board at 6:45 P.M. Struwe was now the deepest diving German, and Lahey had completed a dive he had dreamed about for years. They were the fifth and sixth men to the bottom of the Challenger Deep. "This is the highlight of my career so far," Lahey said.

A half hour later, the errant lander *Skaff* surfaced. The crew attached the lift line and pulled the lander back onto the ship. Vescovo rushed over, hugged the lander—which was quickly becoming his favorite—and then checked the watch. It had stopped after being stationary for two and a half days, but there was no water inside the face.

Hector Salvador removed the watch and handed it to Vescovo. The time read 3:07. Vescovo shook the watch to trigger the self-winding mechanics. Lo and behold, despite being at 16,000 PSI for sixty-two hours, the Omega began ticking as if nothing had happened. Hollywood couldn't have scripted it better.

⁂

On Triton's discretionary dive, Lahey and Ramsay completed a successful dive to the shallower Central Pool of the Challenger Deep on May 5, which made Ramsay the deepest diving Brit in history (breaking the record Jamieson, a Scot, had set in the Java Trench a month earlier). Ramsay, who was uncomfortable during the entire dive, was happy he did the dive, but vowed not to do it again. "No chance," the designer of the sub said.

The final dive on May 7 was the science dive with Vescovo and Jamieson. This would be the first manned dive to the Sirena Deep. Like clockwork, after a four-hour descent they reached the bottom at noon.

But while Vescovo and Jamieson were some 32,000 feet underwater exploring the bottom of Sirena Deep, a crisis of another kind occurred. Buckle, who had emailed the authorities in Guam that the *Pressure Drop* would port on May 8, received a letter from Captain Christopher Chase, the head of the U.S. Coast Guard in Guam, stating that the ship did not have the required permit to enter the port and would not be granted an exemption. The letter further stated the ship's formal application was still being processed in Washington, D.C., which could take another week. If the ship came within twelve nautical miles of the island, the Coast Guard officer wrote, it would be a Class A felony for the captain and the ship would be subject to a fine of $91,000 for each passenger.

McCallum, who seldom gets ruffled and tends to let bad news roll off his back, was irate. "This guy is a small-brained mammal," he said. "As for all their coins that Victor took down to the bottom, we are going to throw them overboard."

The issue was that the *Pressure Drop* had been granted a onetime exemption to port when it docked at the beginning of the voyage. Because McCallum had not received all of the guidelines for porting, he had filed the application too late to receive approval by the time the ship sailed. He had assumed, because of the Coast Guard's VIP reception of the ship, that a further waiver would be granted if the application was not processed by the ship's return date. Indeed, the Coast Guard's office in Washington, D.C., had told him that they would approve the exemption once Guam signed off on it.

McCallum immediately emailed Captain Chase and appealed for him to grant the waiver. "We have 48 souls on board, are immediately offshore Guam and have concluded our Hadal diving program," McCallum wrote. "A week offshore simply waiting for approval in what should be a straightforward administrative process seems an unnecessary waste of time and resources. We have a team of 6 film crew on board who are filming a 5 x 1 hour documentary for the Discovery Channel . . . filling in a week at sea is not going to portray either the USCG or our team in a great light."

Buckle called the Coast Guard office and reached Chase's executive officer, Lt. Commander James Shock, but made no progress. Shock said that the paperwork had to run its course.

Once the *Limiting Factor* was recovered and Vescovo and Jamieson were back on board, Vescovo was called into the captain's office—never a good sign, he had begun to learn—and told about the problem. It was going on 6:00 P.M. in Guam, which was twelve hours ahead of Washington, D.C. What to do?

Vescovo quickly recalled that Dick DeShazo, Caladan's CFO, had recently told him that he had met retired General Charles Krulak, the former commandant of the U.S. Marine Corps, at a Christmas party. Krulak had served as president of Birmingham-Southern College, from which DeShazo had received his master's degree. The two men discussed the Five Deeps Expedition and the general remarked that he would follow its progress. Vescovo asked DeShazo to get in touch with the general, inform him of the record-breaking dives and ask if he knew anyone at the Coast Guard who could help them out.

DeShazo emailed Krulak, who responded immediately that he would be more than happy to help. Vescovo then called Krulak from the ship's telephone, and in very direct fashion, the general asked, "What exactly is the nature of the problem?" Vescovo walked him through all the details of what was going on. When it was obvious that it came down to a piece of routine paperwork not yet being signed in D.C., the general bluntly stated: "This problem will be fixed in forty-eight hours."

Krulak, an undeniable legend in the Navy and Marine Corps community, picked up the phone and called Lt. Commander Shock in the Guam office, identified himself and asked him to allow the ship to port. Powerless without the paperwork, Shock explained that he needed authorization. So, a now equally-annoyed-as-the-expedition-crew Krulak went to work getting it.

"He was on alert that this was going up the chain quickly," Krulak recalls.

Krulak had also served with the deputy director of the Military Office at the White House under President George H. W. Bush, where he had worked on a classified project with Robert Kelly, a former Coast Guard captain, and Steve Flynn, a former commanding officer in the Coast Guard. He emailed Kelly the details, telling him about the Five Deeps Expedition, fully explaining the situation, and asking if he knew anyone at the Coast Guard who could help. "I know how much you like a challenge," Krulak wrote.

Twenty minutes later, Kelly emailed Flynn, who it turns out had been a classmate of Admiral Karl L. Schultz, the current commandant of the U.S. Coast Guard, and forwarded Krulak's email. Flynn promptly emailed Schultz, apprising him of the situation. "I am extremely hesitant to go straight to the top with a request but when Gen. Chuck Krulak, 31st commandant of the United States Marine Corps, who I worked for during my Bush 41 stint, asks via another stint during that time, Capt. Bob Kelly, U.S. Coast Guard, I snap to," Flynn wrote to Schultz, forwarding him the email chain from Krulak. "This seems to me a situation that could and should be very quickly resolved with some encouragement to the command in Guam that it exercise the kind of judgment that distinguishes Coast Guard commissioned officers from the typical federal bureaucrat."

Schultz did just that. Within thirty-six hours of DeShazo's first email—even given the twelve-hour time difference—the *Pressure Drop* received permission to port in Guam.

Krulak was pleased with the result. "Schultz did the right thing thinking, 'This is insanity, Victor Vescovo is a true American hero, forget the paperwork, it will get there, let him come to port,'" Krulak said. "It was fun bringing the band back together and seeing how fast we could get this done."

❧

The *Pressure Drop* ported on the morning of May 9, having achieved a series of firsts in deep diving—and with McCallum holding the bag of coins he threatened to throw overboard. The *Limiting Factor* completed five dives below 10,000 meters. It added four names to the list of people who have been to the bottom of the Challenger Deep. Vescovo broke the depth record, diving on his second solo dive to an announced depth of 10,928 meters (+/- 8 meters), or 35,853 feet (+/- 26 feet), which would later be adjusted to 10,925 meters when all the sonar, sub, and lander data was scrutinized and finalized months later. Walsh and Piccard had gone to a corrected depth of 10,912 meters (+/- 5 meters), while James Cameron had reached a maximum stated depth of 10,908 meters.

The icing on the cake was that Struwe granted full commercial certification to the submersible, the first time any full ocean depth-capable manned submersible has been given such as safety rating by a class agency.

On the science front, the cameras captured better footage of the deepest living fish at about 8,000 meters, *Pseudoliapris swirei*, that Jamieson had filmed two years earlier. They also filmed the ethereal sailfish for the second time ever in the shallower depths of the dive, but again, had clearer footage. There was other interesting footage, but no major discoveries, as the Mariana Trench had been heavily studied by other expeditions' landers.

"We did have great fun, and it *felt* like a big deal," Jamieson said. "I guess I have come to terms with the fact that I couldn't drive this very scientifically, but rather just to do the best I could as opportunities come up and bite my tongue otherwise. But the game changer was when Victor offered me the chance to dive. Rather than another pointless dive to the Challenger Deep, I asked for a dive to Sirena Deep, as we had reasons to believe it was more interesting. Setting down on the seafloor at 10,710 meters was amazing, and in three hours on the seafloor, we actually mapped a hell of a lot of habitat and found more sulfur-based chemosynthetic mounds."

Though the manipulator arm had not functioned well enough for Vescovo to bring back a rock from either the Challenger Deep or the Sirena Deep, the sub had inadvertently captured a rock in the Sirena Deep when the sub bumped a large boulder and a fragment lodged in its battery compartment. Fryer was overjoyed.

"The rock Victor brought back from Sirena Deep is indeed the deepest piece of mantle rock ever recovered from the surface of the western slope of the Mariana Trench, which is the overriding tectonic plate," she said. "It's altered and that alteration may reveal interaction with either seawater or deep-derived fluid from the subducted Pacific tectonic plate. Analysis will tell."

Again, it wasn't a breakthrough scientific find or an aha moment, but one that could begin to reveal new information about the deepest place on Earth. Both Jamieson and Fryer hoped that the images captured by the cameras would help scientists better understand non-photosynthesis-based organic reactions and life development that is outside the norm of regular human experience.

On the final night in the sky bar, Buckle reflected on the dives over beers with Jamieson. "From the beginning of time until about two weeks ago, only three people had been to the bottom of the Challenger Deep," the captain said. "We put four more people down there in 10 days, and everyone is acting like it's not that big of a deal. That's how far we've come from sea trials when the sub couldn't even dive."

Leaving the ship the next morning, Vescovo stopped to reflect on the dives. "The Challenger Deep and the Sirena Deep are actually not deep, dark scary places," he said. "It's just tough to get to them, but once you do, there are amazing things to see."

CHAPTER 16
BREAKING NEWS

Everything was clicking. The expedition was one "deep" away from completing its core mission. Though no one was taking anything for granted, confidence was high that the Molloy Deep in the Arctic Ocean, the shallowest of the five oceans, would be done. Moreover, after the unprecedented technological assault on the Mariana Trench, now was the time to step up marketing efforts to sell the *Limiting Factor*, expand press coverage of all the world firsts, and to dive some new targets.

Coming off the Challenger Deep, there were lingering gripes among the expedition team about continuing restrictions on media outreach. McCallum and Buckle both lamented that part of the problem in getting the U.S. Coast Guard to fast-track the permit to port in Guam was because the expedition hadn't announced its dives to the bottom of the Challenger Deep in real time. Of course, filing the proper permit in a timely manner would have helped even more.

Those gripes would soon be a thing of the past. Geffen, who had not been on the ship, was working with his media team and Atlantic and the Discovery Channel's PR team, in concert with the Richards Group, to prepare a major media campaign trumpeting the news of the dives, with Vescovo coming to New York for a series of print, online, and network TV interviews.

On May 9, after the dives had been completed, a member of Atlantic's media team sent out briefing notes to journalists at the BBC as background material, as the BBC had previously been given an exclusive. This was done to give the BBC a broad outline of what might be coming in the official release, but the BBC journalists were told that in no way was this information for publication and it was directed to use the official, forthcoming press release before publishing any material. The following day, on May 10, there was a conversation prior to the footage of the piece of plastic being sent to the BBC journalists, and in which the Atlantic team reiterated that any report should be held until the final press release.

However, these materials were not cleared by Vescovo or the Richards Group, per the established protocol. Under scientific finds and observations, the notes said: "On Victor's dive he observed at the deepest point of the planet a plastic bag and sweet wrappers." The Atlantic media team followed this up by sending out a picture of the trash at the bottom. A close-up showed what appeared to be a piece of plastic with an "S" on it. In fact, Vescovo hadn't seen any "sweet wrappers," a British-ism for candy wrappers, but he had seen the piece of plastic. The official seven-page press release put out by the Richards Group on May 12, in fact, did not even mention the piece of plastic. Even though the BBC was only supposed to use information contained in the final release, somehow, a disconnect occurred.

On May 13, the BBC ran a report with the headline: "The Mariana Trench: Deepest-ever sub dive finds plastic." That report was picked up by many international news organizations, and most of the press coverage centered on the piece of unidentified trash that Vescovo saw.

The *Washington Post* piece's headline was: "He went where no human had gone before. Our trash had already beaten him there." The lead of CNN's online piece was: "An American undersea explorer has completed what is claimed to be the deepest manned sea dive ever recorded—returning to the surface with the depressing news that there appears to be plastic trash down there."

The trash became the narrative. It was by far the most asked-about aspect of the full ocean depth dives in the TV interviews that Vescovo did. Vescovo was interviewed by Shepard Smith on Fox News. Smith opened the piece standing in front of a screen showing a beach littered with garbage. After running a video of the *Limiting Factor* touching down on the bottom, he zeroed in on the piece of plastic.

A *CBS Evening News* report was headlined "American breaks record with deepest submarine dive ever, finds trash." *CBS This Morning* did a longer, more balanced segment. While the correspondent, Mola Lenghi, asked about other trash Vescovo had seen on his ocean dives, Lenghi also talked to Vescovo about the technology involved in the sub and sea creatures he had seen. Lenghi concluded the piece on a high note, calling the descent to the Challenger Deep "one giant dive for mankind."

While the Five Deeps Expedition was finally receiving national attention, Vescovo was slightly discouraged by the media's focus on trash and that the incorrect "sweet wrappers" line had made it into the BBC notes. He had hoped that the media would place more attention on the not one—but four—dives, getting the first British and German nationals down to the bottom of the Challenger Deep and that the angle of the deepest salvage mission ever to find the lander—which was basically ignored—would have been played up.

"It seems the media can take on a life of its own with a breaking story that has a strong potential hook," Vescovo emailed Geffen. "Just unfortunate that the initial talking points had some inaccuracies that found their way into the main headlines. Maybe I just have a different sense of importance. The media really, *really* liked to bite that biscuit of 'plastic in the oceans.' Even though it was one—and relatively small—piece of trash. Otherwise I didn't see anything like it, and yet it very much dominated the coverage. All I can do is shrug, and make sure that I absolutely do not emphasize it in the future."

Geffen replied that he was "disappointed that, despite being sent all the correct information, the BBC journalists apparently ignored it. It would seem, looking at the output of the BBC Teams unit here, they seem to have an agenda around plastic."

Vescovo also took issue with another citation in the unapproved press notes. The notes stated: "Separately on this mission; Deep Sea Arthropods were found at the bottom of the South Sandwich Trench. The arthropods are big enough to sample fluid and muscle tissue from individual organs containing cilostazol. With the extreme cold conditions and huge pressure at the bottom of the trench, the cilostazol sourced could be a super strand and the volume sourced could help advance the knowledge of how cilostazol works and help further the research into curing Alzheimer's. (Paul Yancey, Hawaii University.)"

Vescovo had cautioned against mentioning any disease cures after the media launch the previous fall when a British paper had alluded to them. Jamieson and the expedition's science team did not agree with Yancey's pronouncements, and in any event, they did not want to speak to individual diseases and certainly not to mention finding cures for them in their press materials.

"I don't think it was an accident that Atlantic—against their contract with me—didn't share the press notes with me prior to their release," Vescovo said. "It's very frustrating, but hopefully it will allow me tighter control in the future."

Geffen disputed that Atlantic had a contractual obligation to share with Caladan an advance copy of background, off-the-record press notes containing nonconfidential information. The real issue was that the BBC journalists apparently did not abide by their agreement to hold the materials as background and only release the information in the official press release. He also rightly pointed out that in the middle of a worldwide conversation about humanity's impact on the environment and the oceans, news of plastic at the ocean's deepest depth would take hold.

"Look, it's news if someone goes to the bottom of the deepest ocean and finds trash," Geffen told Vescovo. "Trash that has made it seven miles down in the ocean is something that can affect our oceans."

On the evening of the 13th, the Explorers Club hosted a special event to announce the record-breaking dives. Board member Ted Janulis introduced the panel comprised of Vescovo, Lahey, Geffen, and Ben Lyons from EYOS Expeditions, attending on behalf of McCallum. He referenced the framed flags hanging throughout the club that marked notable exploration missions, including the ones that went on the Apollo missions. He said that flag No. 81, which he had given Vescovo the previous October to take to the bottom of the oceans, would be returned when the Five Deeps Expedition completed its mission.

After a presentation and panel discussion, the room was opened up to questions. One of them was about the trash found in the oceans. Vescovo repeated that he had seen garbage in the Challenger Deep, the Puerto Rico Trench, and in the Java Trench. But after being asked about it all day, he adopted a more resigned, philosophical approach. "Hopefully, these reports will bring more awareness and more respect for the oceans not becoming a garbage dump," he said.

To dovetail with the worldwide media coverage, Omega launched its ad campaign. On May 15, it ran the first in what would be a series of full page ads in the *New York Times* international edition and other international papers. Featuring shots of Vescovo climbing into the *Limiting Factor*, the sub floating on the surface, and, of course, the new Seamaster Planet Ocean watch, the ad copy read: "Congratulations! Wearing a Seamaster Planet Ocean inside his submersible, Victor Vescovo had defied the crushing pressures of the ocean to reach a new World Record depth of [10,928 meters] (35,853 feet). Adding another chapter to our iconic diving legacy, OMEGA congratulates the entire expedition team on this history-making descent." Vescovo never heard anything, ever again, from Rolex.

∞

A month after completing the Mariana Trench dives, the *Pressure Drop* next ported in Nuku'alofa, Tonga. The ship picked up the expedition team for two planned dives to the Horizon Deep, the nadir of the Tonga Trench named for the vessel that discovered it in 1952. The area was best known as the place where the thermoelectric generator from Apollo 13 splashed down after the 1970 mission was aborted when an oxygen tank exploded. Existing, yet inconsistent, bathymetric data showed that the depth of Horizon Deep was *very* close in depth to the Challenger Deep.

On June 5, Vescovo dived solo and made diving history again by becoming the first person to dive to the deepest point in the Southern Hemisphere, which was measured at 10,823 meters, with a 10-meter uncertainty, or 35,509 feet with a 33-foot uncertainty. The sonar and the CTD data from the sub and landers had definitively confirmed that the Horizon Deep was *just 102 meters shallower* than the Challenger Deep, making it the second deepest spot on the planet.

Though both are in the Pacific Ocean, albeit separated by some 3,700 miles, they were vastly different environments. "The Challenger Deep was like a sandy, undulating beach, calm almost, in its own way," Vescovo said. "But the Horizon Deep had sharp ledges, pebbles all around the bottom, exposed rock that I had to be careful I didn't strike with the sub. And it could be my imagination, but it felt colder."

Because of the broken terrain, the lighting from the sub was also more scattered, adding to the darker and more foreboding feel. During the dive, the *Limiting Factor* had another battery issue, which heightened the tension. This time, water ingress into an electrical junction box caused an electrical "thermal event" and melted the box—though at such depth the fire had no flames.

"I felt like the Horizon Deep did not want me down there poking around," Vescovo said. "It was the eeriest, most hostile place I've seen in all the dives. Partly real, partly imagination, but it really did feel like that feeling you get when the lights go down in a theater and a horror film is about to start, just a general sense of dark, and dread."

The combination of a massive storm the size of New Zealand with thirty-foot seas heading to the dive sight and Triton needing two days to replace the junction box ended the trip with just the one dive—still, the primary mission had been accomplished.

With the press coverage from the *Limiting Factor* logging six dives below 10,000 meters—remarkable considering only two dives by different craft had ever previously been done to that depth—an added push was being made to find a buyer for the Triton Hadal Exploration System, which comprised the submersible, the ship, the support ship, the support boats, the three landers, the ship-mounted sonar, and all spares. The price was set at $48.7 million. (Compare this to NOAA's entire 2018 budget for "ocean exploration and research" of $36.5 million.)

Vescovo and Triton had arrived at that figure based on the actual cash costs to develop, construct, and test the sub and all the other system components over the past four years. It also included reasonable overheads and some production gross margin for Triton that would be expected for any manufacturer, as well as costs related to the sub being commercially certified by DNV-GL. Vescovo would not realize any profit from the sale, and would actually take a fairly substantial loss, as the expedition costs were not included.

Triton's exclusive right to broker a sale of the system had expired in February, which meant that if someone else found the buyer, Triton would have to contribute to the brokerage fee. Bruce Jones had been unsuccessful in drumming up any serious interest. His most promising lead was Rear Admiral Tim Gallaudet, the assistant secretary of commerce for Oceans and Atmosphere and deputy administrator at NOAA. Gallaudet was trying to bring together a government consortium to purchase the system, but that looked like a long shot. Somewhat surprisingly, no other government agency, nor any branch of the military, had shown any interest in the system.

Jones had also done a direct mail campaign of more than 2,600 eight-page color brochures to the super-yacht industry, including captains, owners, naval architects, and shipyards. He emailed more than 2,500 single-family investment offices, pitching legacy opportunities to buy or lease the system. He had also contacted numerous wealthy individuals and more than 280 marine science research organizations worldwide. None of that activity had generated any serious interest.

By far, the best bet was Ray Dalio, one of the highest profile entrepreneurs investing in ocean research. Head of the world's largest hedge fund, Bridgewater Associates, with a personal net worth north of $18 billion, he was profiled by *60 Minutes* on April 7, with correspondent Bill Whitaker touring his luxurious ocean research ship.

Triton had begun talking to Dalio and his team at OceanX in March. The OceanX project was exploring the oceans in the sunlight zone in far shallower waters than the Hadal depths that the *Limiting Factor* was visiting, making the deep-diving sub attractive to the OceanX scientists. Triton had sold Dalio two 3,300/3 submersibles (which could take a pilot and two passengers to 3,300 meters) and worked on several projects with him.

It was beneficial that James Cameron was involved with OceanX and had floated the idea in emails with Vescovo of some kind of partnership between OceanX and Vescovo, or a sale of the Triton Hadal Exploration System to OceanX. McCallum believed that if Cameron asked Dalio to buy the *Limiting Factor*, he would.

McCallum had also presented the system to Australian billionaire Andrew Forrest, who had recently announced a $100 million philanthropic donation for ocean conservation. Forrest had been invited to join the Five Deeps for a dive.

In the wake of the successful full ocean depth dives, OceanX's interest level had increased. Its science and operational team had accepted an invitation to see the sub in action and to go on a dive during the mid-July science mission in the Puerto Rico Trench.

After the Tonga Trench dive, the *Pressure Drop* sailed with its transit crew through the Panama Canal to Puerto Rico, a brutal thirty-day crossing in the wide open, fierce waters of the Southern Pacific Ocean. Triton sent two of its team to Panama to perform routine maintenance and ensure the sub was good to go for its OceanX audition.

On the way to Puerto Rico, just after midnight on July 11, the *Pressure Drop* reached a milestone when the ship crossed its previous track in the Caribbean Sea from November 3. "In September, 1519, Ferdinand Magellan set off on his around the world cruise—the first one in history," Vescovo emailed the expedition team. "And now, almost *exactly* 500 years later, the FDE [Five Deeps Expedition] has gone around—and under—the world's oceans."

~

Three dives were planned for the return trip to the Puerto Rico Trench, a science dive, a dive with OceanX, and a night dive for the film crew. McCallum suggested they do the science dive first without the OceanX team on board and return to port to pick them up—just in case there was an issue.

Vescovo and Jamieson were replicating a dive done by the French submersible *Archimède* in 1964. The French scientists had, very surprisingly, reported seeing schools of fish at a very deep 7,200 meters, though they did not have any video or photographic evidence. The dive plan was to traverse a mile on the bottom in that area and film whatever sea life was there—if indeed there was any.

McCallum's hunch was realized. The dive was aborted at 600 meters because of an outside battery failure—not unlike the failure that had occurred during Vescovo's dive at Horizon Deep—that had caused the sub to lose half its power. The cause: two *tablespoons* of water had seeped into the electrical junction box.

"It was almost certainly there before we launched," Vescovo said. "How the 'F' did it get in there, and why was there no detection of it prior to diving either by the Triton team or in pre-dive checks? At the end of the day, sometimes the Triton guys—all well-meaning—just miss some things when servicing the sub. It should be like an operating room when they change out the oil for these electronic components. They are just not process-driven enough sometimes. It is also a design weakness that water in one junction box can cause us to abort a whole dive. We are working on how to prevent that in the future."

On another headshaking note, the manipulator arm was still not working properly. After the failure of the arm to function—yet again—during the Mariana Trench dives, Vescovo had written about the arm's problems in the blog for the Five Deeps website and mentioned Kraft TeleRobotics as the manufacturer. Lahey emailed Vescovo and asked him to change the blog, as Triton was stepping up and accepting responsibility, rather than deflecting it to Kraft, the manufacturer. The issue was that arm worked fine on deck in the air, but when it went into the ocean, water ingress in the junction box was causing it to misbehave or short out. "It's like the scene in *Dr. Strangelove* when Peter Sellers's hand goes all crazy," Lahey said.

Part of the struggle was that this was just the second full ocean depth–capable arm that Kraft had delivered. The first was fitted to the *Nereus* hybrid ROV formerly owned and operated by WHOI. And part of the difficulty was that there was no way to test the arm at full ocean depth before it was deployed. It could only be tested on an actual mission.

"As you have stated, this is cutting-edge stuff," Lahey replied to Vescovo. "We are testing and using equipment at the hairy edge and as with so many things, it's been an evolution. We build, test, fail, modify, re-test, repair, and then hopefully, we get it."

Lahey had a point, but Vescovo's gripe was also valid—this had been going on for six months. Triton's Frank Lombardo would be attending a five-day school at Kraft that following week to ensure that going forward Triton could fully rebuild any area of the arm at sea, rather than having to ship the arm back to the factory. Lombardo was also purchasing a cache of spare parts to bring to the ship.

After Triton repaired the junction box issue that night, the OceanX team boarded early the following morning. It was led by Dr. Vincent Pieribone, vice chairman of Dalio Ocean Initiative and a professor of Cellular and Molecular Physiology and Neuroscience at Yale University. That afternoon, Lahey and OceanX's chief pilot dived the sub to 3,000 meters. Aside from the manipulator arm not working, the dive came off without a hitch. Pieribone and his colleagues were suitably impressed with the system.

The OceanX team remained on board to observe Vescovo and Jamieson's science dive the following day. On the first attempt, the dive was aborted at 1,500 meters. A drip from the vent on the hatch leaked onto the control console and caused the thrusters to badly misfire, pitching the sub to the left.

It was embarrassing with OceanX watching, and Vescovo was understandably ticked off. Lahey diagnosed the leak, and a quick fix was worked up. At least, Vescovo noted, the rapid fix showed how easy the sub was to fix and how careful they were in conducting sub operations.

Vescovo and Jamieson dived again the next day, and all systems worked as they should. They dived to 7,200 meters, the same as the *Archimède*, but didn't see any of the fish reported by the French scientists. They did see sea cucumbers and other marine life—and yes, human contamination. They spotted a Coke can, a blanket, a porcelain bowl, and a plastic grocery bag. In total, they spent three and a half hours on the bottom and covered over a mile of seafloor.

On the recovery, Vescovo hovered about twenty meters below the surface so that divers could film the sub. After twenty minutes, the divers cleared and Vescovo hit the thrusters to surface. The sub didn't move.

Seconds later, it signaled two battery failures, one on each side. Vescovo had already dropped the freeboard weight that pushes the sub well over the water line.

"It felt like I was stuck at ten meters underwater," he said. "God forbid, we didn't start going down."

Then, for the first time, the oxygen alarm sounded. Jamieson began to feel rattled. Vescovo cranked up the oxygen, part of procedure in such an event, and assured Jamieson everything was fine, that it was just a bad reading. Vescovo rebooted the systems, bringing everything back into norms.

After the *Limiting Factor* was back on deck, Blades figured out that the sub was riding lower on the surface because it was heavier than it had been on Vescovo's solo dives. The sea state had also notched up during the dive, making it even harder to break the surface with the added weight. The solution was to make sure the weight of the passenger was accurately factored in, with a slight underweight modification for possible sea state changes.

That evening, the OceanX team met with Vescovo, McCallum, Buckle, and Lahey. Pieribone said that they were going to recommend the Hadal system to Dalio for a TV series they were doing with the National Geographic Channel. A potential snag, he cautioned, was that rather than owning the system outright, Dalio would likely prefer to have a consortium of owners and divers, possibly including NOAA, WHOI, and Vescovo's nascent research organization, Caladan Oceanic.

Vescovo told them he was fine working the OceanX, but that he didn't want to be in the sub leasing business. He was happy to be a partner in future diving expeditions with OceanX and other organizations, provided that he didn't have to hold the title to the sub and ship.

Pieribone said that after they made the presentation to Dalio, he would try to set up a meeting with Vescovo and Dalio in the late summer to discuss the possibilities. While he did not promise, there was at least a possibility.

~

Around this time, two other potential buyers surfaced. During the Challenger Deep dives, Don Walsh had contacted Sir Richard Branson, a longtime friend, and told him about the *Limiting Factor*.

Branson was a natural candidate. In 2011, he had partnered in the purchase of the *DeepFlight Challenger*, a one-person sub designed by Graham Hawkes for explorer Steve Fossett to dive the Challenger Deep. He announced his Virgin Oceanic project to dive all five oceans, but after numerous problems with the sub during shallow-water sea trials, Branson said that he was putting the project on hold pending a technological advancement in the field.

Well, here it was.

In late June, Vescovo had emailed Branson and invited him to the mid-July science dives in Puerto Rico, or to meet the ship in port in Bermuda in late July when it was en route to the *Titanic*. "It is indeed a pleasure to make your acquaintance, sir, and I have at all times tried to mention to the press that a key seed for this entire undertaking came from you and your idea to execute the Five Dives," Vescovo wrote. "Thank you, most earnestly, for making that first bold attempt."

Branson replied immediately, inquiring as to the sale price and the operating costs. Vescovo told him the price and detailed the rationale behind it. He revealed to Branson that the vessel costs were several hundred thousand dollars per month at full manning while running operations. Branson expressed interest, emailing that he would do his best to pay the ship a visit in Puerto Rico.

The second buyer that had surfaced was someone who professed to have Department of Defense contacts. He had emailed Triton with his interest. The man had a very limited digital footprint, but there was evidence that he had done covert work for the U.S. government in the 1980s. To find out if the guy was real and had the resources, Lipton, Vescovo's attorney, was going to start by asking him to place $5 million in a trust account.

The inquiry was interesting because there had always been a feeling that the U.S. government would, in some way, be interested in at least having use of the sub. Perhaps this person was a vehicle for different government agencies to fund and share the use of the sub without dealing with the normal red tape of naval appropriations. "Big Navy sourcing moves at the speed of a drunken mastodon," Vescovo only half-joked. "Maybe this is a way around it."

Vescovo contacted his college roommate, Enrique Alvarez, who was an FBI cybercrimes investigator. Under federal regulations, the Hadal system would qualify as a dual use system—meaning it involved counter proliferation for advanced technology. Therefore, Vescovo would have the right to request an FBI report on a potential buyer.

Days of intrigue and batting around James Bond–like scenarios came to a crashing halt after Lipton vetted the candidate. Regardless of his contacts, he simply didn't have the funds to pursue the transaction, and some of the documents he sent to Vescovo just didn't look "right." He had also not gone public with his interest, so it was a mystery as to what he had to gain by expressing interest in pursuing a deal he couldn't afford. In a huff, the buyer said he would just source the sub from another manufacturer and buy another, shallower-capable sub as well. Over a year later, he had not surfaced again in the submersible industry.

❧

On July 15, the *Pressure Drop* left Puerto Rico bound for Bermuda, where it would pick up the expedition team for a second attempt to dive the *Titanic* wreck and have Atlantic Productions film it for a one-hour special coproduction for National Geographic Channel and other broadcasters. That afternoon, Branson emailed Vescovo that he was putting his interest in the Hadal system on hold. He had just announced plans to take his space-tourism venture, Virgin Galactic, public by the end of the year, making it the first listed company that planned to take tourists into space.

"Have thought long and hard on this one and discussed with my team and the general feeling is that whilst we are so involved in the space projects that we should concentrate on those at the moment," Branson wrote. "We are going through a funding programme for those and if successful we would be happy to revert back in 6 months' time if you haven't sold it by then. I think it is better to be upfront on this and not waste your time."

Yet again, the sexy space race appeared to have trumped ocean exploration. Vescovo and Lahey were obviously disappointed. While the focus for an onward sale would now be squarely on OceanX, they both planned on doing that thing that only they could do—keep on diving and showing just what the *Limiting Factor* was capable of. They remained focused on what else could be observed, studied, or discovered. Diving on the *Titanic*, the most famous wreck in the world, could do just that.

CHAPTER 17
BACK TO THE *TITANIC*

Diving the *Titanic* wreck was not part of the Five Deeps Expedition's core mission, but Vescovo had decided to return for a second attempt in late July. From a practical standpoint, he often remarked that it was "on the way" to the Arctic Ocean, the fifth deep, so "Why not?" But exploring the wreck was also about enhancing the profile of the *Limiting Factor*, almost a noblesse oblige for the world's first formally accredited full ocean depth manned submersible.

The fact of the matter was, no matter how impressive the sub's record dives were, the bottom of the oceans can feel remote and esoteric to people who had heard and read news reports of the dives. By diving the *Titanic* wreck, the *Limiting Factor* could add to the living history of the tragedy and conduct forensic work on the condition of the wreck, which had not been visited in a manned submersible since 2005. The one-hour special that Atlantic Productions was making for National Geographic Channel and others broadcasters globally—which was set apart from the five-part Discovery Channel series on the expedition—ensured a high profile media platform for the mission.

The story of the RMS *Titanic* has become one of legend and lore that has continued to attract interest over time. The details of the voyage remain startling: the largest man-made moving object in its day sailed with 2,206 passengers of all social classes and crew on its maiden voyage. Four days into its passage from Southampton, England, to New York on April 14, 1912, the ship hit an iceberg, split in half, and sank to the bottom of the Atlantic Ocean several hours later, resulting in more than 1,500 deaths, including many well-known people.

James Cameron's 1997 film put the audience on the ship and connected people to the tragic voyage in a way that had never been done. The immense popularity of the film—it won a record-tying eleven Academy Awards and became the highest-grossing film ever—sparked a renewed interest in the ill-fated voyage and the marine archeology of the wreck.

The themes of *Titanic*'s story still resonate: the bravery of those who sacrificed their lives to save others, the cowardice of some who boarded lifeboats ahead of women and children stranded in the bowels of the ship in third class, the heart-rending dedication of the band that played on as the ship sank, the nobility of the wealthy men who donned tuxedos and sipped brandy as the ship carried them to their deaths, and over it all the arrogant folly of mankind, who believed it could build a vessel that "God himself could not sink." People project their own behavior onto those whose stories have been told, and often ask the question of themselves: would such a life-or-death tragedy have brought out your better angels or your darker demons?

"There is something in this story that grips people," explained Parks Stephenson. An expert on all that is *Titanic*, he has spent more than twenty years studying every aspect of the wreck in an attempt to piece together the full story of the voyage. "Everyone finds something in this disaster that connects to them. It's so all-encompassing."

~

Spirits on the *Pressure Drop* were high when the ship sailed from Bermuda bound for the *Titanic* wreck on the sun-kissed afternoon of July 25. In addition to the expedition's assault on the Challenger Deep and its successful audition for OceanX, the weather forecast for diving the *Titanic* looked very favorable. A year earlier, the expedition had spent five days at the wreck's site and not completed a single dive due the perfect storm of a massive hurricane to the south heading for the site, a weather front bearing down from the north, and strong currents coming across from the west.

During the five-day transit, the expedition spent time preparing for the dives. Stephenson had constructed intricate eight-inch models of the bow and stern of the ship so that everyone could visualize the wreck. Based on the most recent images from a Woods Hole Oceanographic Institution ROV dive in 2010, a laser-generated printout was combined with the models to make a diorama in the control room that the sub pilots and surface team could use for reference.

Five dives were planned, the first a solo dive by Vescovo. Both Lahey and McCallum were concerned that a solo dive was too dangerous because of the strong currents and lack of a navigator, but Vescovo was insistent. No one had ever done a solo dive to the *Titanic* before, and he wanted to be the first. When pressed to reconsider, he just shrugged.

Unfortunately, Lori Johnston, the microbial biologist who had left the last scientific plate to detect the rate of decay, had gotten sick before the trip. She was the only one who could collect, treat, and fully assess the rusticles. (Rusticles are the complex microbiological structures that feed on the ship's iron, forming into the shape of icicles on the surface of the wreck. The word was coined by Robert Ballard, who discovered the *Titanic* wreck in 1985, to describe the rusty-looking icicles dangling from the sunken ship.)

Not having the key scientist on board was something of a letdown, but the wreck could still be assessed from the cameras' footage. The scientific plate could be retrieved for examination in the lab and a new plate could be placed.

This would be the first time that the *Titanic* would be filmed with high-resolution cameras that would produce 4K images. The cameras were the key to examining the wreck in greater detail. Past expeditions had grappled with how to film dynamic images of the wreck. Even with powerful exterior lights on the subs and ROVs, the video frames were heavily pixilated, and could only be corrected so much by an editor. This expedition's goal was dramatically cleaner, crisper images.

Vescovo had been in touch with James Cameron via email about Cameron's previous dives. Cameron cautioned him to be really careful, warning "it's a dangerous place," and told him that the sub's sonar would be his best friend and navigational guiding light.

The wreck lies 2.4 miles underwater in a canyon adjacent to the Continental Shelf, which creates a swirling current running up to two knots that can make it look like there is blizzard on the bottom. With limited visibility, the only way to determine positioning until the final meters would be the sonar. The primary danger was that the *Limiting Factor* would have to come so close to the wreck to film it that the sub risked being pushed into the wreck by the current and possibly entangled in its rope and wire cables, which still dangled from its various masts.

Triton made a rescue plan in the event the sub became entangled. There were four vessels in the area carrying ROVs that could be called upon and sent down to the wreck to aid the *Limiting Factor* in the event of a distress call.

The required court order from the U.S. District Court in Norfolk, Virginia, specifically stated that the sub could not make contact with the wreck or the seafloor. This strict provision had been instituted because previous expeditions had deliberately cut into the metal so that ROVs could fit through and film the inside. The expedition was not allowed to remove any artifacts. However, it was permitted to take no more than 500 grams of rusticles off of the wreck. To ensure procedures were followed, an NOAA representative was on board, as was P. H. Nargeolet, who had been on thirty dives to the wreck and was acting on behalf of the RMST, the organization charged with recovering and preserving artifacts from the wreck.

The expedition was also authorized by the court to remove scientific plates that had been placed there from 1998 to 2004. These plates, which attract the rusticles from the ship, would show the rate of deterioration of the wreck. A new plate would be placed to measure future deterioration.

Stephenson, who worked on Cameron's 2005 *Titanic* expedition, was hoping for new finds that would help tease out more of the story of the ship and its passengers. On his wish list were images of the captain's quarters, the gymnasium, and the never-before-seen bulkhead "K." He also wanted to find the stateroom of the industrialist Benjamin Guggenheim, the richest man on the *Titanic*, which had been ripped off Deck "A" and was somewhere in the debris field.

"My worst fear is that there has been massive collapse of the wreck," said Stephenson, as he assessed the previous images. "If that is the case, every trip there is critical because the wreck won't be with us much longer. But we will see."

⁂

On the first dive, with Vescovo piloting solo, the *Limiting Factor* reached the bottom at roughly 3,800 meters, or 12,500 feet, in a quick ninety minutes, twice as fast as the *Mir* that Cameron dived. The very strong current, however, pulled the sub about 500 meters away from its target point. Vescovo hit the thrusters, and while closely monitoring the sub's sonar, headed for the wreck.

For twenty minutes, he didn't see anything on the sonar, but he knew he was close. Then, there was a prominent sonar return that looked almost like a cliff in front of him. He crept up to it slowly. The 1.5-knot current in the canyon was creating a vortex that was moving him at will. Alternating between the vertical set of thrusters and the horizontal ones, he fought against the current to maintain a steady, straight pace.

As he moved closer to the "cliff," the seafloor became more rigid and a berm of sand the size of a large wave appeared through the bottom viewport. The sonar indicated that the wreck was fifteen meters in front of him, but all he could see through the viewport was the sand berm.

And then there it was—a sheer wall of black-painted iron protruding upward from the seafloor some twenty to thirty meters high: the bow of the *Titanic*.

"Oh my gosh," he said in amazed tone of voice that the sub's interior camera picked up, "I'm at the bow . . . I'm at the bow of the *Titanic*."

For someone who had said a year earlier that his personal interest level in seeing the ship was a 3 on a scale of 1 to 10, he was actually awestruck. He realized that the berm had been made when the ship smashed into the seafloor. The wreck was *huge*.

"I didn't realize how big the ship was," Vescovo later said. "I was like an ant looking up at a human. It was tough fighting the current while trying to get good footage. To be honest, I couldn't just sit back and savor the experience. I felt like a fighter pilot, always on, wondering where am I, where's the current, what's behind me, what's in front. It was nerve-wracking and exhausting."

Cameron had told Vescovo in an email that he had gotten stuck at the bow on one of his dives and needed a couple of hours to work free. Vescovo now clearly understood what he was talking about. This place, he was constantly reminded, was dangerous.

He drove around the bow for ninety minutes in what was a virtual cloud of particulate matter. Even though the wreck was a massive structure, Vescovo's visibility was limited to ten to fifteen meters, so he hung back a bit. He could clearly see the rusticles that were hanging from the handrails. They looked like gold-colored icicles, and some were five feet long. He could see that the square windows of the "A" deck were all enveloped by corrosion, which was actually the bacteria eating away at the wreck and leaving behind waste product.

He next headed for the stern, which was 600 meters away. This was easier said than done given the poor visibility, the current, and the sonar's effective range of about 200 meters maximum. As planned, he positioned the sub at the aft end of the bow section, "aimed" the sub in the direction of the stern, and set out at maximum thrust to minimize current drift. As he traveled along at a fair clip, he saw a seafloor littered with bowls and plates, intact bottles, crockery, and bathroom tiles. He even spotted a perfectly preserved par of trousers.

"The whole area was very jumbled and chaotic, like a war zone," he later said. "There was twisted metal and cables everywhere."

After about twenty-five minutes, he received a substantial return on the sonar indicating that he was near the stern. He slowly eased back on his thrusters until it came into better view. Looking out the viewport, he saw huge beams with what looked like hooks on them, or "dragons' teeth" he later called them. To him, they were intimidating as hell. These were part of the understructure that was peeled backward when the ship slammed into the seafloor.

For the next half hour, Vescovo navigated—extremely carefully—his way around the stern.

"I was sweating bullets most of the time, trying to get good footage," he said of the experience. "You are either too close or not close enough. If you are too far away, you get no visuals. You need to be within ten meters, but that is dangerous because metal and God knows what else is jutting out to the side and below you."

After more than three hours exploring the wreck, he was mentally exhausted from fighting the current and worrying about hitting the wreck or becoming entangled. He thrusted away from the wreck to a designated point where he could drop his weights without contaminating the wreck, let the surface know he was coming up, and surfaced without incident.

As exhilarating as the dive was, there was a comedown when the film crew discovered that the cameras on the front of the sub, including the camera to capture the 4K images, had not worked. Although there was no forward footage, the downward high-definition cameras had functioned and delivered some arresting images. The issue was diagnosed and a firmware patch was downloaded so that it would not happen again. Vescovo sighed when he heard the news and commented: "Well, I've learned these things never go right the first time out."

Stephenson studied the footage for changes from the images on previous dives. He observed that part of "A" deck had collapsed. From his detailed knowledge, he was able to relate what was happening on the ship in the areas filmed and bring them to life. Pointing to a davit on the video screen, he explained: "Right in this area, first officer Murdoch was trying to load the last lifeboat into the davit so it was launched properly. This davit is cranked in such a way that shows that Murdoch was nearly done when the crew was literally swept away by the ocean." Those were the stories that could still be told by examining the wreck.

❧

The day before the second dive, Stephenson pored over the footage of Ballard's 2004 expedition. He was searching for the location of the early scientific plates. Because he knew the wreck intimately, he had a pretty good idea of where to look.

Stephenson knew that Ballard had laid an experiment and a plaque on the floor next to the stern. But because an earlier experiment was already there, Ballard had changed his mind and tried to lay the plaque on a rusticle pile next to the starboard stem, where it sank into the pile. He moved his experiment to the floor outside of the mud piled up near the portside deck. He then picked up an earlier experiment from the boat deck and relocated it near where he had placed his experiment.

"We knew they were taken off the ship by Ballard, but we didn't know where they were," Stephenson said. "I found the sequence of photos where Ballard lifted them off the deck and put them on the ocean floor."

The second dive would attempt to retrieve the scientific tray and place a new one close to the wreck so that rusticles could latch on. With Lahey piloting and Magee, who had previously been to the wreck, navigating, the two made it to the lander on the starboard side of the wreck without too much difficulty.

Using the manipulator arm, Magee grabbed the rope attached to the tray. He lifted it out, off of the lander. Lahey then drove to the wreck, where Magee successfully dropped the tray. The two high-fived. "How about that!" Lahey said.

Lahey then navigated to the port side to recover the tray that Stephenson had located. Using the manipulator arm, Magee grabbed the tray, and Lahey drove to the lander. When the sub reached the lander, Magee lowered the tray into the basket. But it turned out that the plate was just slightly larger than the basket. He precariously draped it over the edge as best he could. However, on the lander's ascent, the tray fell off.

It was unfortunate. The expedition was now going to have to base its study of the deterioration on the footage they captured on the sub's cameras and not on hard physical evidence.

On the third dive, the expedition team determined the best combination to get the most complete footage of both the bow and the stern was for Lahey, the most experienced sub pilot on the ship, to pilot the sub and for Vescovo to navigate.

With now-practiced efficiency, the duo quickly reached the bow and filmed it for almost two hours. Their primary objective complete, Lahey retraced the path to the stern and once again navigated its massive piles of destruction and debris. Lahey was bowled over by the size of the wreckage and its dilapidated condition. "It's a rusty pile of junk!" Lahey exclaimed. Vescovo admonished him: "Stop saying that, Patrick, we're on camera."

Shortly after surfacing, it was clear that this time the cameras worked, and thus, this three-hour dive was the most successful yet in bringing back detailed footage of both the bow and the stern. The first 4K video images of the *Titanic* had finally been secured.

Stephenson went on the fourth dive, with Vescovo piloting. He had been down to the wreck once before, but felt that he needed to see it again. In addition to continuing to capture images of the gymnasium and the captain's quarters, the dive's other goal was to find the Guggenheim stateroom that had been sheared off near the stern. Finding the remains of the relatively small stateroom was a big ask in a debris field that stretched fifteen square miles.

On the dive, Vescovo first took Stephenson to the bow section. As they filmed around the captain's quarters, Stephenson was glued to the viewport. "It's devastated," he said. "I see a glimpse of the [captain's] tub, but it's disappearing."

Again, the currents were swirling, and at times the thrusters were fighting something of a losing battle. At one point when Vescovo was moving sideways along the right side of the bow, where it dropped down into the forward cargo area, the current pushed the *Limiting Factor* hard against its directional wishes, and the sub bumped the wreck.

Vescovo immediately blasted the thrusters to pull away and up. Both he and Stephenson felt the impact, but Vescovo didn't think that the impact had damaged the sub. All systems were still working and he had a "green" board. Later analysis of external video showed that the sub had actually bumped a small piece of railing that was jutting out at a 90-degree angle from the ship that did not show on sonar, and was not visible from the cockpit.

After another half hour of surveying the bow, perhaps even *more* carefully and slowly than before, he pulled back and headed across the debris field to search for the detached remains of the Guggenheim stateroom. The current was so strong that the sub ended up far to the east of where they thought they were. Vescovo and Stephenson ended up approaching the stern from the northeast, rather than the northwest. It took a bit of navigation and investigative fieldwork to figure out where they were, but they eventually managed to get near where they thought the Guggenheim stateroom might be.

Soon, Stephenson spotted some metal ribbing on the bottom, with some unique characteristics—at least to Stephenson's trained eye.

"This might be Guggenheim," he said, looking out the bottom viewport.

Vescovo maneuvered the sub as close to the seafloor as he could.

"That's it! That's it!" Stephenson said. "This is the Guggenheim stateroom right down there." In the vast multitude of the entire wreck of the *Titanic*, they had managed to find a specific stateroom. They high-fived each other.

Back on the deck, the *Limiting Factor* was inspected for damage. It had a small, rust-colored spot on it from the contact with the wreck, but no real damage to speak of. P. H. Nargeolet considered it so minor that he didn't even mention it to RMST. Still, McCallum would have to address the sub bumping into the wreck in his written report to NOAA and the U.S. District Court for the Eastern District of Virginia.

McCallum wrote that the expedition had taken extreme care in trying to preserve the wreck and abide by the rules of the permit. However, he wrote, because of an extremely strong and unpredictable current, the sub did make incidental and accidental contact with the seafloor and on one occasion with the ship. "There was no evidence of visible damage to the submersible," he wrote. Both he and Vescovo griped to each other that they were surprised that their permit had not even allowed for "accidental" contact, but that language had somehow been missed during permitting. Nargeolet confided to Vescovo later: "Honestly, your expedition probably did far less damage to the wreck than any of the others."

On the final dive, Lahey and Frank Lombardo, who was in charge of the manipulator arm, dived the sub to try and work out the problems with the arm, which would operate for brief periods and then fail. Even after the latest fix to solve the hydraulic pressure issue and the time that Lombardo spent at Kraft TeleRobotics learning all aspects of its functionality, the arm continued to malfunction after just a few minutes of use. It was becoming like an incurable allergy, or a Dr. Strangelove moment, as Lahey called it. Unfortunately, the team was having to complete the research and development of the arm in situ, in the ocean and in the deep pressure environment that they couldn't replicate on land.

Though the "hard" science component of collecting the tray had not been successful, the five dives filmed the wreck as never before. By performing dedicated photogrammetry passes on the wreck, the most accurate 3D models of the *Titanic* yet were produced by Atlantic's film crew. In addition to helping assess the wreck's current condition and project its future, these made it possible to visualize the wreck using Augmented Reality (AR) and Virtual Reality (VR) technology.

Johnson, the microbial biologist, later examined the footage to gauge the level of deterioration. She was pleased that a new, highly visible plate had been placed for bacteria to populate over the course of a number of years so that it could later be retrieved and correlated with the wreck itself.

"The microbial bacteria has become the dominant organism," she said. "Honestly, there is more life on *Titanic* now than when she was floating on the surface."

The detailed images captured by the *Limiting Factor* would continue to fill out a story that will never be completely told. After examining all of the footage and making comparisons to his library of images, Stephenson outlined places that were disappearing.

"The most shocking area of deterioration was the starboard side of the officer's quarters, where the captain's quarters were," Stephenson said. "The captain's bath tub is a favorite image among the *Titanic* enthusiasts, and that's now gone. That whole deck hole on that side is collapsing, taking with it the staterooms, and the deterioration is going to continue to advance."

Yet, at the same time, Stephenson saw the possibility of new avenues opening as future explorers would be better able to see inside of the ship.

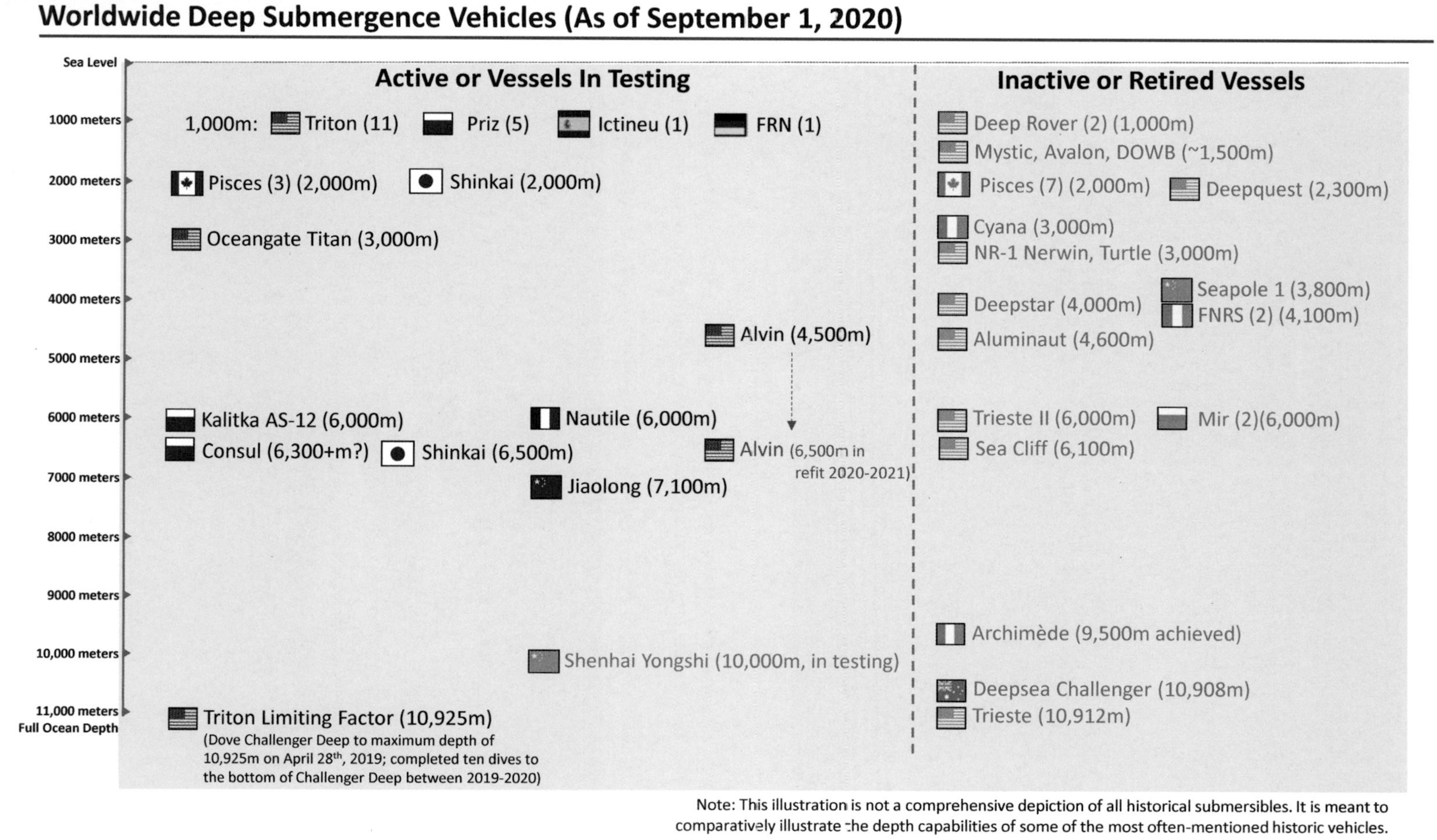

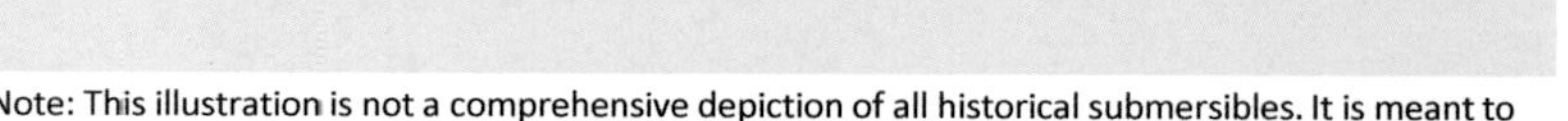

Fig. 1: A Snapshot of the History of the World's Deep Ocean Submersibles. *Credit: Caladan Oceanic.*

Fig. 2, ABOVE: The titanium pressure hull of the *Limiting Factor. Credit: Josh Young.*
Fig. 3, BELOW: The *Pressure Drop* in port prior to sailing around the world. *Credit: Josh Young.*

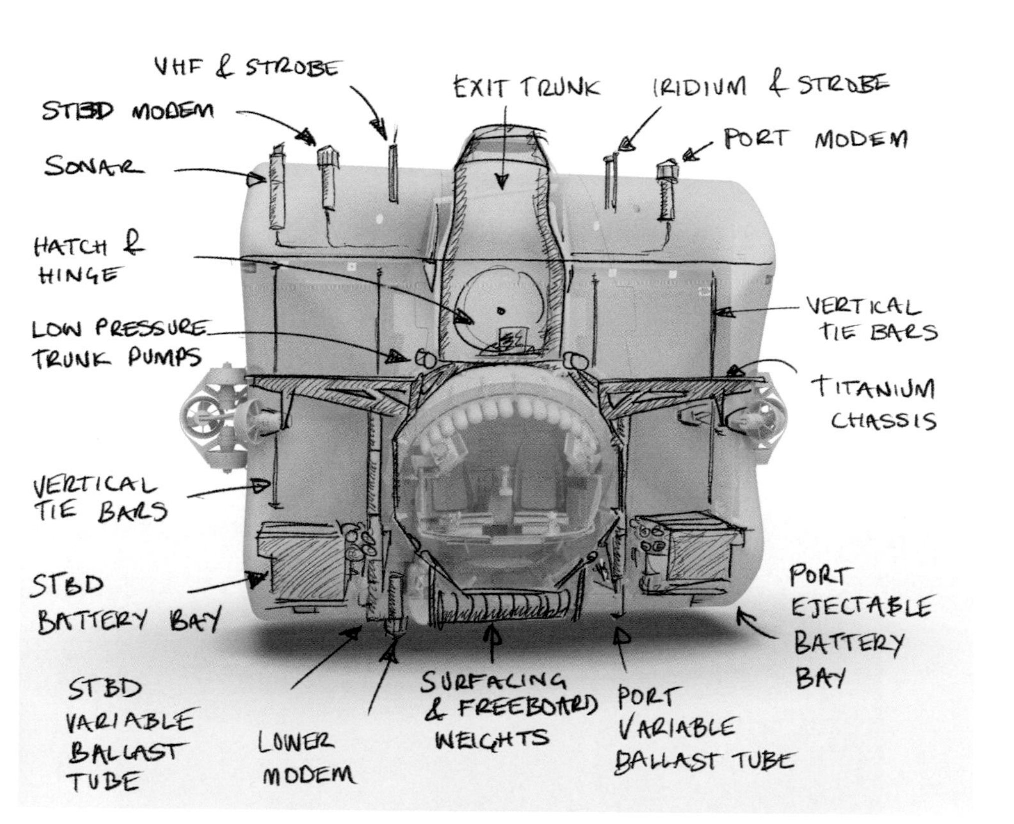

Fig. 4: Sub designer John Ramsay's sketch of the *Limiting Factor. Credit: Triton Submarines.*

Fig. 5, ABOVE: Victor Vescovo, who funded the Five Deeps Expedition and served as chief pilot of the *Limiting Factor*'s dives to the bottom of all five oceans. *Credit: Tamara Stubbs.*
Fig. 6, BELOW: Five Deeps Expedition leader Rob McCallum (foreground) planning a dive with Dr. Glenn Singleman (background). *Credit: Reeve Jolliffe.*

Fig. 7, ABOVE: Triton Submarines co-founder and CEO L. Bruce Jones. *Credit: Mindy Miller/ Triton Submarines.* **Fig. 8,** BELOW: Triton Submarines co-founder and president Patrick Lahey. *Credit: Reeve Jolliffe.*

Fig. 9, RIGHT: The Five Deeps Expedition team at the Explorers Club in New York announcing their mission. From left to right: Caladan Oceanic technical consultant P. H. Nargeolet, Triton's Patrick Lahey, Victor Vescovo, expedition leader Rob McCallum, chief scientist Dr. Alan Jamieson, Atlantic Productions CEO Anthony Geffen. *Credit: Caladan Oceanic.*

Fig. 10, LEFT: Presenting at the International Robotics Showcase in London. From left to right: Triton's electrical systems design engineer Tom Blades, sub designer John Ramsay, Triton co-founder Patrick Lahey (holding a viewport from the *Limiting Factor*), Anthony Geffen, executive producer of the five-part Discovery Channel series on the expedition. *Credit: Josh Young.*

Fig. 11, ABOVE: The *Limiting Factor* hanging from the A-Frame crane, ready to launch with Triton team member Frank Lombardo in the foreground. *Credit: Reeve Jolliffe.*
Fig. 12, BELOW: Planning the first dive in the Southern Ocean. From left to right: P. H. Nargeolet, geologist Heather Stewart, Dr. Alan Jamieson, Patrick Lahey, Victor Vescovo, *Pressure Drop* captain Stuart Buckle, chief sonar operator Cassie Bongiovanni. *Credit: Atlantic Productions.*

Fig. 13, ABOVE: *Pressure Drop* relief captain Alan Dankool on the ship's bridge. *Credit: Enrique Alvarez.* **Fig. 14, BELOW:** The *Pressure Drop* in King Edward Cove, Grytviken, South Georgia Island, where it stopped on its way to the Southern Ocean. *Credit: Atlantic Productions.*

Fig. 15: *Skaff*, one of three full ocean depth-capable scientific landers, being deployed. *Credit: Reeve Jolliffe.*

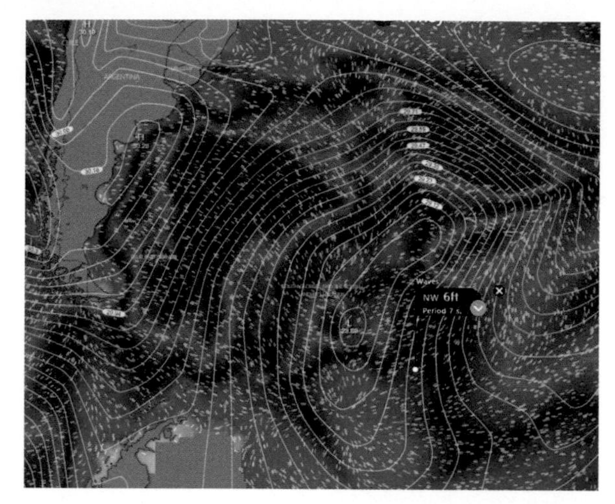

Fig. 16, LEFT: "Here Be Dragons": A map of storm systems (purple areas) in the "screaming sixties," the area in the Southern Ocean known for its ferocious weather. *Credit: Caladan Oceanic/ Screenshot from windy.com.*

Fig. 17, RIGHT: The *Limiting Factor* in King Edward Cove, South Georgia Island, preparing for a test dive. *Credit: Atlantic Productions.*

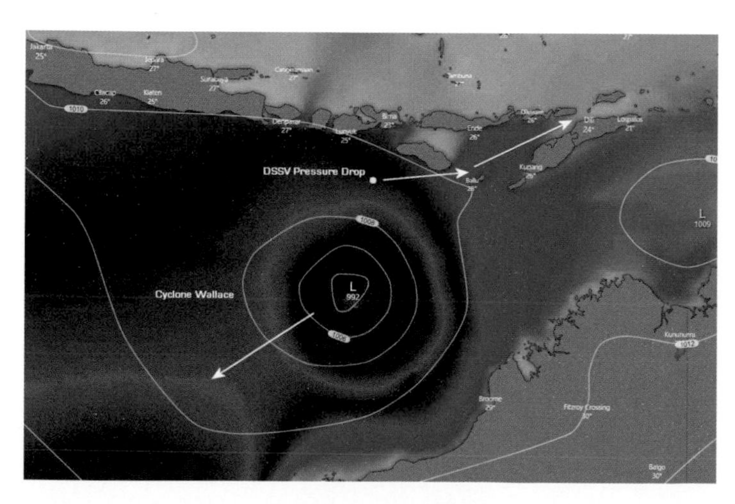

Fig. 18, LEFT: The *Pressure Drop* plotting a course to steer clear of Cyclone Wallace in the Indian Ocean en route to Dili, East Timor. *Credit: Caladan Oceanic.*

Fig. 19: Victor Vescovo paying tribute to the legendary explorer Sir Ernest Shackleton at his gravesite in Grytviken, the abandoned whaling station on South Georgia Island. *Credit: Caladan Oceanic.*

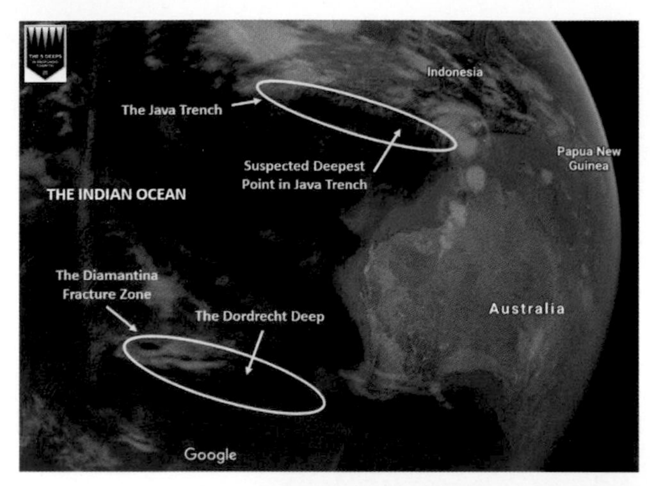

Fig. 20, LEFT: The deepest parts of the Indian Ocean: the Java Trench, mapped and verified as the deepest point by the expedition team at 7,192 meters (23,596 feet), and the Diamantina Fracture Zone, mapped slightly shallower at 7,019 meters (23,028 feet). *Credit: Caladan Oceanic/base map screenshot from Google Maps.*

Fig. 21, RIGHT: A new type of species discovered in the Java Trench that was first dubbed "Snoop Dog" by expedition scientist Alan Jamieson, who later concluded it is a tunicate sea squirt (*Ascidian*). *Credit: Atlantic Productions.*

Fig. 22, LEFT: The deepest ever dumbo octopus (*Grimpoteuthis*) filmed in the Indian Ocean's Java Trench. *Credit: Atlantic Productions.*

Fig. 23, ABOVE: The Five Deeps Expedition team at the Challenger Deep. From left to right, back: EYOS Expeditions's Rob McCallum, Victor Vescovo, *Pressure Drop* captain Stuart Buckle, chief scientist Dr. Alan Jamieson, Triton Submarines co-founder Patrick Lahey, and, in front, expedition guest U.S. Navy captain Dr. Don Walsh, commander of the *Trieste*, the first submersible to reach the bottom of the Challenger Deep. *Credit: Atlantic Productions.* **Fig. 24,** BELOW: Australian "swimmer" and submersible technician Tim Macdonald, who had the harrowing job of releasing the *Limiting Factor* from the ship to dive and then resecuring it to the ship on the surface. *Credit: Enrique Alvarez.*

Fig. 25, ABOVE: Victor Vescovo boards the *Limiting Factor* on the deepest dive ever to the bottom of the Challenger Deep in the Pacific Ocean. *Credit: Reeve Jolliffe.* **Fig. 26,** BELOW: Victor Vescovo piloting the *Limiting Factor* on the way down to the deepest place on Earth. *Credit: Atlantic Productions.*

Fig. 27, ABOVE: The scientific platform *Skaff*, commonly called a "lander," at the absolute bottom of the deepest ocean collecting data. *Credit: Caladan Oceanic.* **Fig. 28, BELOW:** The expedition's sonar-generated dive map of all three pools (Western, Central, and Eastern) of the Challenger Deep, showing all dives made there by the end of 2019. *Credit: Caladan Oceanic.*

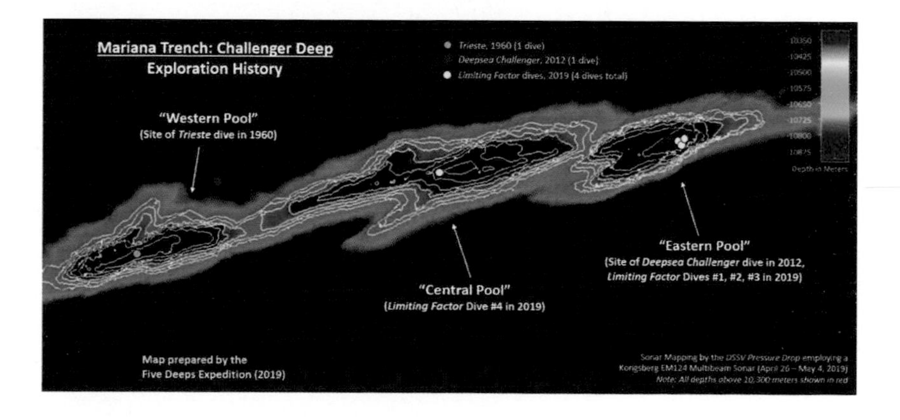

Fig. 29, LEFT: Microbial mat at the bottom of the Sirena Deep in the Mariana Trench, taken from inside the *Limiting Factor*'s pilot capsule. *Credit: Caladan Oceanic.*

Fig. 30, RIGHT: A piece of suspected plastic at the bottom of the Challenger Deep photographed by Victor Vescovo through the viewport of the *Limiting Factor*. *Credit: Caladan Oceanic.*

Fig. 31, ABOVE: The rocky edge of the Challenger Deep in its Eastern Pool. *Credit: Caladan Oceanic.*
Fig. 32, LEFT: *Hirondellea gigas*, a deep sea amphipod that lives at the bottom of the Challenger Deep. *Credit: Dr. Alan Jamieson.*

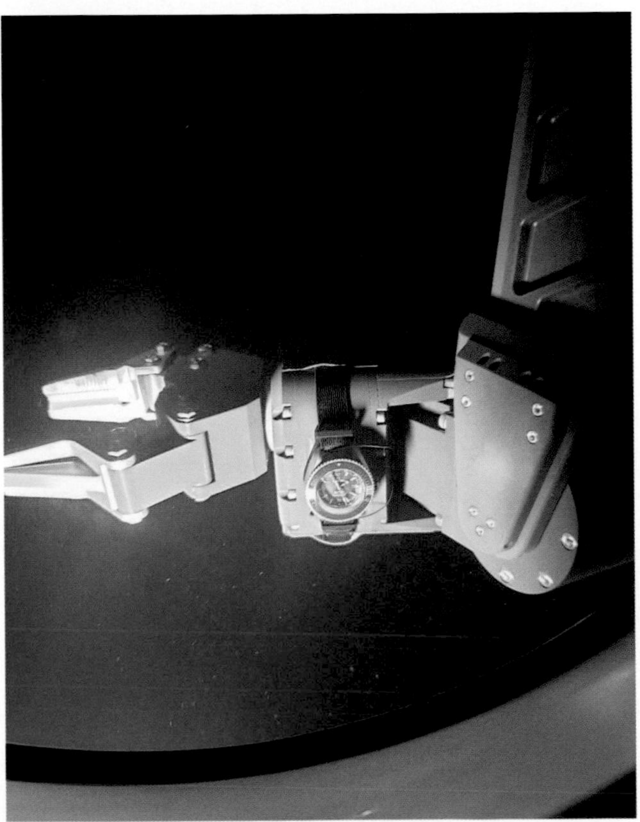

Fig. 33, ABOVE: The deepest place on Earth, the bottom of the Challenger Deep in the Pacific Ocean. *Credit: Caladan Oceanic.* **Fig. 34, LEFT:** The Omega Seamaster Planet Ocean Ultra Deep Professional timepiece strapped to the *Limiting Factor*'s manipulator arm on the way to full ocean depth, for its first of seven journeys to the bottom (as of September 2020). *Credit: Caladan Oceanic.*

Fig. 35, ABOVE: Dive planning in the Arctic Ocean. From left to right: Caladan Oceanic technical consultant P. H. Nargeolet, Triton's Patrick Lahey, *Limiting Factor* pilot Victor Vescovo, expedition leader Rob McCallum. *Credit: Enrique Alvarez.* **Fig. 36, BELOW:** Recovering the *Limiting Factor* after its record-breaking dive at the Challenger Deep. *Credit: Reeve Jolliffe.*

Fig. 37, **ABOVE:** The first mantle rock recovered from the Sirena Deep in the Pacific Ocean, at approximately 10,900 meters. *Credit: Atlantic Productions.* **Fig. 38,** **BELOW:** Victor Vescovo signals the crew after diving the Molloy Deep: "Five!" he shouted standing on the hatch. *Credit: Atlantic Productions.*

Fig. 39, ABOVE: Victor Vescovo on the *Pressure Drop* in the Arctic Ocean. *Credit: Enrique Alvarez.*
Fig. 40, BELOW: The arctic ice pack several miles away from the Arctic Ocean's Molloy Hole. *Credit: Enrique Alvarez.*

Fig. 41, ABOVE: Remnants of the Grand Staircase at the wreck of the *Titanic*. *Credit: Atlantic Productions.* **Fig. 42,** BELOW: Victor Vescovo and French rear admiral Jean-Louis Barbier dove to the wreckage of the downed French submarine *Minerve* at the bottom of the Mediterranean Sea in an attempt to determine how it sank. *Credit: Caladan Oceanic.*

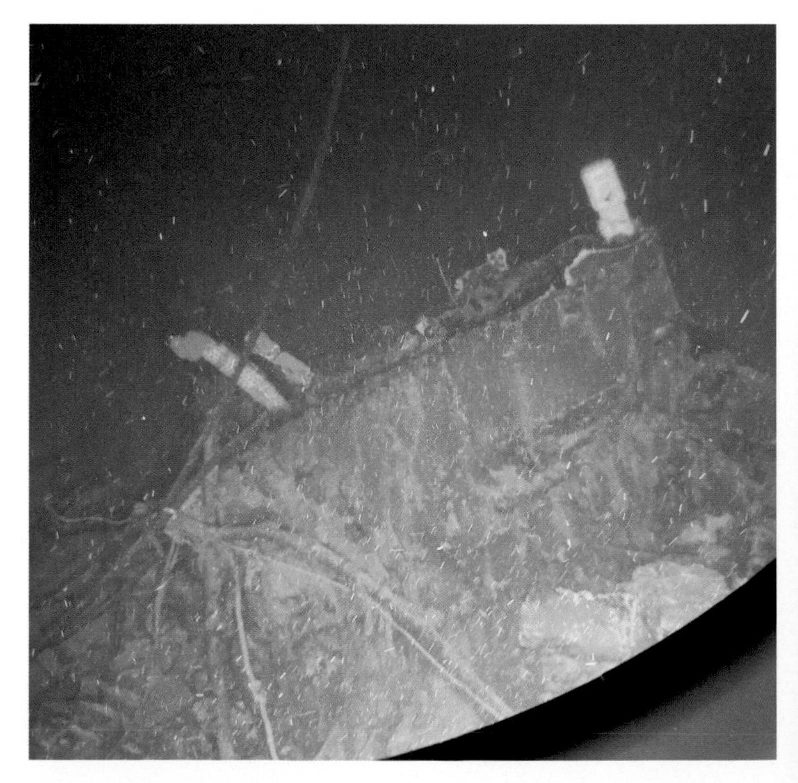

Fig. 43, ABOVE: The edge of an underwater volcano in the Suakin Trough in the Red Sea. *Credit: Caladan Oceanic.* **Fig. 44, LEFT:** HSH Prince Albert II of Monaco, an international ocean advocate, and Victor Vescovo after diving the Calypso Deep. Prince Albert became the deepest-diving Head of State with the dive. *Credit: Caladan Oceanic.*

Fig. 45, ABOVE: Victor Vescovo and Dr. Kathy Sullivan, the first American woman to conduct a spacewalk and the first person to ever visit space and dive to the bottom of Challenger Deep. *Credit: Caladan Oceanic.* **Fig. 46,** BELOW: Victor Vescovo toasting the success of the Five Deeps Expedition at the Royal Geographical Society in London with members of the expedition team, guests, and family members. *Credit: Reeve Jolliffe.*

"It's a shipwreck that's returning to nature," he said. "But that doesn't mean that we are done exploring *Titanic* yet. This deterioration is opening up new possibilities to learn more about the wreck. We went down there to look, to document, and to assess. The result of this expedition wildly exceeded our expectations."

Initial headlines in October about the expedition screamed that the wreck might "disappear entirely" in a few years. Vescovo and Stephenson agreed that this was just a lot of hype, quoting a single scientist with an extreme—and likely incorrect—view to garner clicks and lure readers. Sure, some places were deteriorating, but it was a wreck at sea. Vescovo commented to his team: "That wreck will be there a hundred years from now, trust me. In much worse shape and collapsed a lot more, sure, but geezus, I don't think people talking about it disappearing soon realize how massive it is."

⁂

On January 28, 2020, several months after the initial stories covering the team's *Titanic* expedition, news of the *Limiting Factor* bumping the *Titanic* broke in the press. A UK *Telegraph* headline blared: "EXCLUSIVE: Wreck of the *Titanic* was hit by a submarine but U.S. government kept it quiet, court told." The article was picked up by several publications, including the *Washington Post* and FoxNews.com.

In the *Telegraph* article, RMST, the *Titanic*'s sole authorized salvage company, alleged that McCallum's company, EYOS, and the National Oceanic and Atmospheric Administration (NOAA), which is responsible for protecting the wreck site, did not inform the court or RMST of the *Limiting Factor* bumping the wreck for five months. The implication was that there was some attempt to cover it up. However, that wasn't the case.

In responding to RMST's allegation, NOAA pointed out that RMST's own observer on board, P. H. Nargeolet, knew about the incident. NOAA said that after the organization reviewed McCallum's report, it reminded him that he was required to provide the report to the Virginia court. When McCallum found that out in early January 2020, he filed the report. Because so much time had elapsed, RMST cried foul.

"A full report was sent to NOAA upon completion of the expedition as was a condition of permissions to dive the wreck," McCallum said in a statement. "It was not, as has been misreported, withheld or delayed."

It turned out that the expedition was caught in the crosshairs of an intense fight in the background between RMST and NOAA over the stewardship of the wreck. Specifically, RMST wanted to go in with several ROVs, cut into the structure of the vessel, and remove the famed Marconi wireless set used for communication to enhance their exhibit in Las Vegas. NOAA was trying to prevent them from doing so, because actually cutting into the wreck was not, NOAA believed, within RMST's salvage rights, would damage the wreck, and disrespect those who died there. It was the "salvage and preserve" versus the "do not disturb/graveyard" faction of the *Titanic* community. Emotions ran high on both sides, with the expedition team caught in the crossfire.

RMST was actually trying to imply that NOAA had improperly supervised the dives and not only allowed for a "collision" with the wreck, but subsequently tried to hide the fact. Vescovo believed they were deliberately inflaming the situation with inaccurate information—proverbial "fake news" he called it—potentially to discredit NOAA and persuade the court to grant them sole discretion to do what they wished at the wreck. As a businessman, he understood why they were doing it, but didn't appreciate their tactics or even the argument itself.

McCallum was having none of it, and went public to say so. In responding to the RMST allegations in a statement emailed to Fox News, he directly voiced concern over RMST's plans: "I am alarmed by the plans of RMST who are seeking permission to breach the hull to remove further items for commercial gain. They have already removed thousands of items from the debris field around the wreck and are now seeking to enter the wreck itself to take more."

In its court filing, RMST also demanded that EYOS turn over *all* footage of all the dives to the court. EYOS, of course, did not have the footage, the UK-based Atlantic Productions did. Rather than let the situation escalate and risk damaging the credibility of the Five Deeps Expedition or resulting in the Walt Disney–owned National Geographic Channel not airing the one-hour special in the U.S., Vescovo asked

Geffen for the clip of the bump so that he could show it at a previously scheduled court hearing in February. There, he reasoned, he could provide the whole context of the incident and replay it for RMST, the judge, and any media that attended.

"The bump just wasn't that bad," Vescovo said. "RMST was making a mountain out of a molehill, and we needed that to stop."

Vescovo voluntarily attended the court hearing on February 20, 2020, and brought the footage of the bump with him. After an extensive explanation of the sub's systems, current conditions, and visibility limitations during the dive in question, Vescovo played the clip in open court. When it was over, the judge cast him a look that to Vescovo, was like a comment of "That's *it*?"

None of the reporters present filed stories on the hearing. The court later indicated that it was not going to pursue "the bump" further. When asked by the judge, the NOAA and RMST lawyers said they were satisfied and would not pursue it further. So much for the "collision" and "cover-up."

Even so, the incident impacted the marketing undertaken by the National Geographic Channel for the special. In an apparent attempt to steer clear of any controversy regarding the incident or RMST's allegations, the program titled "Back to the *Titanic*" aired on February 23, 2020, in the United States virtually without any promotion, a disappointment considering all the work and expense that had gone into the two trips to film the wreck. The show was preceded by a rerun of the 2017 program "*Titanic*: 20 years later with James Cameron." "Back to the *Titanic*" managed a 0.22 rating, just shy of the 0.24 that Cameron's show garnered.

After watching "Back to the *Titanic*," Vescovo recalled seeing the wreck through the viewports of the sub for the first time and the wave of amazement that swept over him, causing him to feel like a ten-year-old-boy seeing a mythical image.

"That's how I will remember *Titanic*," he said. "If our contribution to the story creates that feeling in others who saw the show or will eventually see it, then I can say, 'mission accomplished.' But I have to say, there are two secrets of *Titanic* I learned. The first secret is that it's *really* big. The second is that it's *very* dangerous"—and here he paused for a roll of the eyes and a shake of the head—"and let's not even talk about the politics involved in diving it."

CHAPTER 18
ROLLING THROUGH THE DEEPS

One year later than planned and three weeks after the successful *Titanic* dives, the expedition team gathered in Longyearbyen on August 21, 2019. This was a quintessential seaman's town if ever there was one, right down to its rustic pubs with the distinction of being the most northern in the world. With a declared population of just 2,368, the settlement in frigid Svalbard, Norway, is best known as the jumping-off point for trips to the North Pole and expeditions to the Arctic Ocean. Most who visit it can't help but liken it to "North of the Wall," an accurate comparison to the northern wasteland in HBO's series *Game of Thrones*.

Vescovo himself had been there two previous times, to train for his Antarctica expedition and as the penultimate departure point for skiing the last 100 kilometers to the North Pole as part of the Explorers' Grand Slam. Now he was returning in an attempt to become the first person to complete submersible dives to the bottom of all five oceans—something there was no catchy name for, though as Lahey quipped, "Whatever they call it, the word 'epic' should be included."

Their destination was the Molloy Deep, or Molloy Hole as it is scientifically known, which is located 160 kilometers west of Svalbard in the freezing waters of Arctic Ocean. Rather than a trench like the deeps in the other four oceans, Molloy is actually a large depression in the ocean floor that is formed where the Eurasian and North American tectonic plates split apart, a process that is ongoing. It is also by far the shallowest of the five ocean "deeps," at 5,551 meters (18,212 feet). Still, only four other active submersibles on the planet could go down this far.

What a long way the expedition had come in a year. Twelve months ago, when technical hiccups caused it to arrive in extreme weather, the expedition had spent five days at the *Titanic* wreck and failed to dive it, and not even attempted to sail north to the Molloy Deep. Now, the expedition team was riding high after completing dives to the bottom of four oceans, yet, in what had become a theme of the expedition, nothing ever came easy.

During the run from Bermuda to the *Titanic* and then onward from Newfoundland to Longyearbyen, the *Pressure Drop* experienced recurring issues with its starboard propeller shaft, which was operating at about 60 percent of full power. Paddy Russell, the ship's experienced chief engineer, detected a noise coming from the shaft during the long transits. This hadn't slowed the ship much, as the gondola producing the "Venturi effect" had increased the ship's speed slightly since its installation in December of 2018. However, there was a cause for concern, as the shaft could deteriorate, or fail completely.

Russell's supposition was that the noise occurred when the ship was lower on fuel, causing the shaft to "ride high" and scrape. A full load of fuel appeared to "bend" the metal and make the shaft be seated tighter, and thus noise free. There was no way to inspect the shaft and determine what the problem was without the ship going to dry dock, as it was located inside a lubricated oil tank on the underside of the ship. He estimated that the ship would need nine days in dry dock for the repair. The process involved removing the shaft, draining the entire system of oil, replacing the damaged bearings and aligning the new prop. This could be done in Newfoundland after the *Titanic* dives, but that would risk missing the Arctic Ocean weather window for a second time, meaning the dive would have to wait yet another year.

Not a chance. The schedule was tight, as usual. The *Pressure Drop* needed to complete the Molloy Deep dives and make it to London for a presentation of the expedition at the Royal Geographical Society set for September 9. The plan was that the ship would then sail on to Las Palmas in the Canary Islands, where this repair and other maintenance would be performed. The shaft had run for 4,000 hours since the problem was first identified, but Russell was uncertain the prop would make it to the finish line. In a slightly unsettling email, he told Vescovo: "I cannot tell you if this will last until dry dock."

In consultation with Captain Buckle and Russell, both highly experienced sailors, the decision was made to go for it. Even if the starboard propeller shaft reduced to an operating rate of below 60 percent, having the port shaft running at full power would allow the ship to complete the mission and make the London date. However, if the starboard shaft trended significantly downward, or stopped working altogether, the ship's speed would be cut in half or more, below what would be considered a safe operating rate on open water.

"All we have to do is just make one more long run," Vescovo said. It was, as usual, a calculated risk.

The *Pressure Drop* did indeed make it to Longyearbyen without incident and set sail from the port on the afternoon of August 22. The weather window that the expedition had missed the previous year looked like it would be open for a few days, so three dives were planned on consecutive days: Vescovo's solo dive, followed by two science dives. When the ship reached the site, it would undertake a quick mapping exercise to find the deepest point. Fortunately, it was a relatively small area so the survey would take less than twelve hours. Time was of the essence because of a massive storm front on its way three days hence.

A slight wave of "contented melancholy" was in the air among the team that had been on the expedition the entire time. They had been through so much, and now they were just one dive from achieving a goal that at times over the past year had seemed like wishful thinking. Some energy was provided by new faces for this leg: a *New Yorker* reporter; world-renowned photojournalist Paolo Pellegrin, who had covered everything from the funeral of Pope John Paul II to fashion shows in New York; a *Wired* reporter; and Vescovo's college roommate and close friend, Enrique Alvarez, an FBI cybercrimes agent and former U.S. Naval Reserve officer who was a talented photographer in his own right.

During the voyage, Alvarez cornered Lahey and expressed concern about Triton's computers being hacked and the plans for the *Limiting Factor* being stolen by a hostile nation or bad actor. Given that it was the only manned submersible in the world that could reach full ocean depth meant it could do quite a bit of harm in the wrong hands, such as cutting international transmission cables. "All cables cut, all financial markets down," as Vescovo had taken to saying. No foreign nation had contacted Vescovo about purchasing the sub, but, the thinking went, if they could steal the technology, why pay for it? Alvarez put Lahey in touch with the FBI cybercrimes office in Miami so that Triton's computer system could be monitored.

"The FBI visited us and made it clear to us that they felt like the Chinese were a serious threat to tech companies like ours," Lahey said. "We were keenly aware of that. At Victor's insistence early on we had strengthened our firewalls and other cybersecurity measures."

Conditions in the Arctic Ocean for the first dive on August 24 were cold and bleak but eminently diveable. The big dangers were a windchill of minus 8 degrees Celsius and the water temperature was minus 2 degrees Celsius, giving anyone who fell into the ocean only a couple minutes before hypothermia gripped them. The other "hazard," as McCallum had started to point after so many successful dives, was complacency.

Despite the harsh chill and a peppy sea, the launch came off like clockwork. The LARS team had hit their stride.

Vescovo was cleared to descend. As he began pumping in, he said, "Last of the Five Deeps . . . such a long journey."

"There she goes," Lahey said from the deck as he watched the *Limiting Factor* disappear from sight.

Vescovo checked in every quarter hour. The dive was going as smoothly as any had. Lahey took to pumping his fists at each 1,000-meter interval. The only issues in the sub were minor—a few drops of water coming from the hatch vent early on had stopped when it sealed with the increasing pressure, and the faint cracking sounds of the exterior as the sub adjusted to the pressure, to which Vescovo had grown accustomed.

As he approached the bottom Vescovo became a little tense. *So close*, he thought. "One hundred and forty-three meters from the bottom," he said to himself, looking at the sonar. "Oh my goodness, I should start to see it very soon." He began to release the small, 5-kilogram VBT weights one at a time to slow his descent and allow him to decelerate into a hover by the time he reached bottom.

A few minutes passed, and there it was, a sight that was growing more familiar to him, this one yet again a sight that no one had ever seen: the bottom of the Arctic Ocean.

"Surface, this is the *LF*," he said. "The *LF* has landed." He paused, then couldn't help himself but add: ". . . That's one small dive for a man, one giant leap for ocean exploration."

The control room engaged in a rousing round of applause and high-fives. "Ha, that's great," Lahey said. "But he's not going to talk anymore. He's going to head to *Closp*."

Closp, the lander, had been dropped at the deepest point of the Molloy Deep. Although both the lander and submersible had been deployed over the deepest point identified by the sonar, the landers seemed to drop a bit more directly down because of their shape, so to officially dive to the deepest point, Vescovo needed to reach the lander. In his mind, the mission wouldn't be complete—not only diving to the bottom of all five oceans, but to the *deepest* point of each one—until he reached the lander. He also believed that just in the act of finding the lander on the seafloor, he would traverse enough ground to significantly increase the odds that he had reached the deepest point identified by sonar.

Soon, his instruments told him he was only 200 meters away. As he moved across the seafloor, he began having navigational issues. He was receiving wide variations in his heading, causing him to wonder if the magnetic field this far north was affecting the navigation system. (It was.) The cold temperature was also beginning to drain his batteries at a faster than normal rate. It was a race against time and magnetic interference at the bottom of the Arctic.

Nearly three hours into his time at the bottom, half of his comms went out—just as they had in the Southern Ocean. Lahey could hear Vescovo, but Vescovo wasn't able to hear Lahey, who issued one of his now familiar "what the fuck is wrong with this thing" declarations. Vescovo ignored the problem and continued his search.

Then, finally, Vescovo saw a "hard" return on his sonar at about seventy-five meters. It had to be it. When Vescovo was twenty-three meters away, he spotted the lander's single, bright light. "I got ya," he said. "There it is . . . the end of a dream . . . I have to touch down."

Two meters in front of *Closp*, he did.

"Surface, this is the *LF*," came the slightly garbled communication from the sub. "Life support good. *LF* has landed at *Closp*."

"Roger that, *LF*," Lahey said. "Understand you are at the bottom. Congratulations . . ."

In the sub, time had slowed down. The explorer had found his grail. It was the feeling of the sprinter breaking through the tape on a gold medal run, the mountain climber reaching the summit, the astronaut stepping onto the Moon. The difference, of course, was that those had all been done by several people. This achievement had only been accomplished by one.

He sat there, quiet and alone in uncharted territory, quite literally, and peered through the viewports, taking in the results of the quest one last time. He slowly pirouetted the sub around, an effortless movement. The pilot and his deep sea mechanical companion had developed their own personalities to each other over the expedition, and they understood one another. The pilot had grown to regard the sub as an idiosyncratic, robotic pet that often acted in its own way, like a colleague that might not always take direction but one that responded symbiotically when necessary. It was obvious to all, and himself, that he adored his machines.

The mission done, Vescovo let out a long sigh, smiled, quietly chuckled to himself. He paused deep in thought about all it had taken to reach this point, took in a few more slow breaths, and finally checked in with the surface:

"Five Deeps done . . . let's go home."

And with that, he dropped his surfacing weights and began his return.

Two and a half uneventful hours later, the *Limiting Factor* breached the surface. Effortlessly, the LARS team put it under tow. When the sub was secured at the aft of the ship, the swimmer opened the hatch.

Lahey was grinning. "This is a big moment, guys," he said. His grin widened. He couldn't help himself.

The hatch clear, Vescovo climbed out and raised his right hand high with all five fingers extended. Everyone on deck joined in the applause. Flares and firecrackers greeted him as he exited the hatch.

"Five!" he shouted. "Five!"

Grasping the hook that secured the sub with his left hand, he stood in place holding up the five fingers, flashing the biggest smile anyone had ever seen from him.

Back on deck, Vescovo hugged each member of the team, offering his sincere thanks.

"What is there to say, I just dove to the bottom of all five of the world's oceans . . . it feels fantastic . . . it's a hell of a thing," he said, pausing to lean into his next thought. "What little kid *doesn't* want to be an explorer. If you put your mind to it and get the right people, almost anything is possible."

His final depth in the Arctic Ocean was recorded as 5,551 meters (18,212 feet) +/- 14 meters. Once again, Wikipedia's entries on the Arctic Ocean and all existing maps would need to be updated.

Dive days spiked the energy level on the ship, and the last voyage was no different. An all-hands morning meeting to discuss and plan the second dive in the Molloy was held at 7:00 A.M. in the control room. McCallum provided the details, and then Lahey weighed in, followed by Vescovo. At the August 25 meeting, McCallum chose not to dwell on the core mission having been completed, and instead insisted everyone keep their focus for the two "icing on the cake" science dives on back-to-back days. Three dives in three days at a deep would also be a first for the expedition.

On the first post-solo dive, Vescovo descended with Scottish geologist Heather Stewart. A marine geologist at the British Geological

Survey, her primary focus was the geomorphology (the shape of the seafloor) and its geological features. The goal of the dive—her first ever in a submersible—was for her to personally see and for the sub to film suspected towers of pale colored carbonate vents in the Molloy Hole.

When they reached 500 meters, however, the hatch started dripping. Stewart looked over at Vescovo for reassurance.

"Is that normal?" she asked a bit nervously.

He stared at it for a solid fifteen seconds and then announced: "Not really," he replied. "But it's okay for right now."

Flashing an "if you're good, I'm good" look, she said, "Okay."

Vescovo filmed the drip at intervals with his camera phone. As the *Limiting Factor* continued to descend, he could see that the drip from the hatch was becoming more aggressive, despite the fact that the pressure should be tightening the seal. It was apparent to Vescovo that something very small was "stuck" in the hatch seal—probably a metal shaving or something similar, preventing it from fully sealing. Frustrated but realistic and safety-conscious—especially with a mother of small girls on board—he called for an abort of the dive, and the LARS team had the sub back on the ship in record time.

Stewart was disappointed, but the crew was even more disappointed than she was. They knew how excited she was to dive. Her first scheduled dive in the Tonga Trench had been scrapped because of weather. And now, her second dive had been aborted. When asked about it, she pursed her lips and managed a smile. "The way it goes, I guess," she said.

On deck, there was frustration, but not the kind of hand-wringing that had occurred earlier in the expedition when dives were aborted. The Triton team knew the sub far better. One of them declared that they simply had to get it fixed so Stewart could dive. Magee quickly determined that the grommet in the hatch seal, which they had replaced just before the dive, was ever so slightly kinked.

Rather than losing the entire day, with only one diving day left, Magee went to work fixing the grommet so that the sub could be redeployed. While he machined the shaft in the ship's metalworking shop and put a new grommet in, the LARS team reloaded the VBT, freeboard, and surfacing weights. Because the launch and recovery had become so smooth, they would be able to dive again within just two hours. Determination and experience had conquered disappointment.

On schedule, the sub, with Vescovo and Stewart back in it, was once again on its way down. This descent was without incident and soon, Stewart saw the bottom of the ocean for the first time, calling it "a truly remarkable experience." They drove around several large ridges, recording the rocky terrain littered with fissures and quite massive boulders. Unfortunately, they did not find the "white towers" of rare carbonate hydrothermal vents they had hoped to see. However, *not* seeing them—like anything in science—provided its own unique data set for future geologists to ponder.

The next day, the final dive of the expedition—along with the final lander deployments—was a shallower dive to survey marine life. Vescovo took chief scientist Jamieson with him. Once again, Jamieson was able to see marine life in action that he had previously only seen on film. Given the shallower depth, the sub filmed, and the landers retrieved, quite a bounty of specimens to examine later.

When the *Limiting Factor* surfaced, however, the control room had a slight problem with the tracking. There were two intersecting points on circle plots where the sub could have been. Usually, it was obvious which location was the correct one. The ship and the recovery boats picked what they believed was the right one and headed there. However, the sub unexpectedly surfaced at the other one, leaving it too far away for the VHF radio signal to reach the ship.

Alvarez, Victor's close friend, went to the bridge with binoculars to scan the ocean for the sub. When he didn't see the sub in the area where the recovery team was positioned off the port side of the ship, he moved to the starboard side. About 400 meters away, he spotted the orange flag on the sub bobbing on the horizon. He alerted Buckle, who radioed the recovery boats and turned the ship around for the pickup.

The sub bounced around for ten minutes before Vescovo reached the recovery boats on the VHF radio. When he finally heard them, he answered their radio queries asking if he was okay by playing "Space Oddity" by David Bowie from his mobile phone, loud enough for them to hear it over the mic: *"Ground control to Major Tom . . . your circuit's dead, there's something wrong . . . Can you hear me, Major Tom? . . . Here am I, floating 'round my tin can . . .*"

As the song played while the boats headed to recover the sub, Jamieson, who had had so many highs and lows with the science part of the expedition, sat back, taking the situation in. In his typical sardonic way, he said, "This is the last dive of the expedition and they've lost us." And then he let out a burst of laughter.

The *Limiting Factor* was put under tow for the last time in the Five Deeps Expedition, 386 days from the first time it was recovered, way back during sea trials. Magee, running deck operations as usual, had a huge smile on his face as the sub was moved into place for Vescovo and Jamieson to disembark. Everyone knew their jobs so well that conversation during recovery was now minimal.

Tim Macdonald, the expert swimmer from Australia, opened the hatch. Jamieson climbed out, followed by Vescovo. Everyone not in a boat or on the bridge had gathered on deck to applaud, from the guest observers to the kitchen staff.

After it was over, Alvarez commented to Magee how smoothly the process went. Magee smiled. "That might be the last time, so we had to go out on a high note," he said, his voice tinged with sadness.

⁂

Immediately after the sub's recovery, the Captain made a course change of 90 degrees and headed for the safety of the northern Svalbardian fjords. Racing to catch the ship was a rather intense polar storm, but it was a race the storm would lose. Since the team had time available before they all returned home, the ship sailed in and out of several fjords, where the weather was calm, the wildlife plentiful, and the scenery majestic. McCallum, who had led several expeditions in the Arctic Ocean, directed Buckle to specific areas where they could see polar bears, walruses, and artic foxes. On the final day before returning to Svalbard, the ship ported on a remote island with an abandoned World War II–era German weather station that had been blown up by Norwegian commandos.

Jamieson and Stewart had begun assessing what had turned out to be a very productive trip at the Molloy Deep. On the very last lander deployment of the expedition—the 103rd—Jamieson filled nineteen jars with samples. He pulled in the biggest amphipod hauls of the entire expedition, mostly a *Eurythenes* species, which resembles a balled-up shrimp. Most thrilling, he had caught a number of Glacial eelpouts, a true Arctic species. "Never seen anything quite like it," he remarked.

On the final night, there was one last cookout in the sky bar, with many bottles of Vescovo's favorite champagne, Veuve Cliquot, consumed. McCallum led the ceremonies wearing his "bad taste" Hawaiian shirt, as he did at every barbecue. There were speeches and toasts from all the team leaders. Everyone was working their way through different feelings on the final night of the expedition.

Vescovo again thanked everyone. "I love you all," he said as he pointed around to the assembled group. "I trusted you guys with my life. You went out on this journey with me and no one got hurt."

John Ramsay, who was now the first man to design a submersible that had gone to the bottom of all five oceans, repeatedly, and ridden in it, too, grew very emotional talking about the expedition. Every dive for him had been a nail-biter, as he waited for something to go wrong and hoped things would go right. Looking back over what he had been through and what had been achieved, he was overcome with emotion. He sat down, put his head in his hands, and began to cry.

~

On September 9, the *Pressure Drop* sailed up the Thames River in London and docked at Canary Wharf surrounded by towering skyscrapers, marking the official completion of the Five Deeps Expedition. En route from Svalbard, the ship had stopped in Edinburgh to pick up Dick DeShazo and his wife, Sandra, and Matt Lipton and his wife, Andrea. Vescovo wanted the two men, who had worked so hard and not gotten to sail on the ship, to have the opportunity to experience it firsthand. Vescovo also boarded there so he could be on board when the ship sailed into London, as did Don Walsh.

The ship had completed a helical circumnavigation of the globe, traveling some 46,000 nautical miles. The *Limiting Factor*, secure in its cradle on the aft deck of the ship, had completed thirty-nine dives, setting depth records in all five oceans. Don Walsh, who piloted the first sub to dive the Challenger Deep, summed it up as "the most ambitious exploration expedition of the century."

Now, it was time to celebrate and share the results of the expedition.

That night, the team leaders all gathered for a cocktail party and program on the expedition at the Royal Geographical Society, one of the premier organizations for the advancement of geographical sciences in the world. Many of their family members attended, as did select media and guests.

For his long-term girlfriend Monika, Vescovo was able to give her a fifth and final gold Cartier bracelet—one for each of the five deeps. Each one was uniquely inscribed with the location of each deep and its depth. It was the only collection of its kind in history.

When Victor's sister, Victoria, entered with her husband and Victor's step-nieces, he darted through the crowd and gave her a hug. It had been a long way from the dinner when he told his sister of the idea and sketched out the logo on the back of a business card.

Discussing Victor's dives, Victoria recalled that during the three hours he lost communications with the ship while diving the Southern Ocean, she had an odd feeling in her stomach and felt an ache in her back. "I knew that the titanium that keeps my spine in place was keeping him safe," she said.

The presentation in the historic ballroom pretty much encapsulated the expedition. The technical challenges and scientific finds were celebrated, the obstacles overcome were joked about, and the lingering personal issues surfaced. When McCallum, who introduced each speaker, came to Geffen (who had not been on the ship for any of the deep ocean dives or the second trip to the *Titanic*), he did not say his name but merely walked off the stage. Jamieson also slighted Atlantic Productions, whom he very much didn't care for, saying that he could not show any footage or they would sue him. Vescovo was disappointed that differences couldn't be put aside after so much had been accomplished, but these were very strong personalities, he figured. They wouldn't be so good at what they did if they weren't. He just grinned and bore it.

But there were certainly moments of levity and humor. Buckle joked that after the 6,000-mile transit from Tonga to the Panama Canal "the fuel light was on." Jamieson, dubbed "Dr. Fucking Jamieson," (whose middle initial was, indeed, "F") was in typical form. When a line on the video screen appeared that said thirty-six new species of crustaceans had been discovered, he hemmed and hawed, "That number is purely random . . . it's plus or minus thirty-five." Of the sub Vescovo quipped: "You can't order this on Amazon."

McCallum called the expedition by far the longest and most complex of the 1,200 expeditions he had ever led and talked about adaptability being a key part of the expedition. He noted that he had made ninety-four mission plans and dealt with fifty-five governmental agencies around the world. And he touched on how the struggles strengthened the team.

"We had to learn by doing, which is a euphemism for learn by breaking," McCallum said. "You start by breaking equipment and then you break people. At the other end, if you can outlast all of the failures, you will find success."

He introduced Lahey, who emphasized what a global undertaking the Five Deeps Expedition had been in every regard. Suppliers for the sub's components had come from Australia, Canada, the UK, Germany, Spain, and the U.S. The hull had been pressure-tested in Russia, and the sonar had been installed in Curaçao. The expedition team had nationals from all over the world, which is why Vescovo had put the UN flag on the sub. The captain was Scottish, as was the chief scientist. The cook was Austrian, and his sous chef was from Argentina. The hotel staff was Filipino. Triton's team had a Canadian leader, a Spaniard, and several Americans. The doctor was Australian. The expedition leader was a New Zealander. The certifier was German, and the film's executive producer and crew were British.

As usual, Lahey wore his emotions on his sleeve. It had been the ride of lifetime, with peaks that felt like mountain tops where he could see forever and valleys that felt like, well, the depths the sub dived where there was only darkness. "I am so proud to have been part of the expedition," he said. "Being part of this historic undertaking has been the greatest privilege of my professional life. I can't say this enough . . . it took a maverick, it took somebody who had an audacious dream that was willing to take a chance on a small, obscure company and give us an opportunity to build a revolutionary craft, the most extreme deep ocean exploration tool in history."

Ramsay, introduced by Lahey as "the Leonardo da Vinci of sub designers," walked through how he conceived the *Limiting Factor* in layman's terms. "We had to keep them breathing, keep them dry, get them down, allow them to see out," he said, as he showed a moving graphic of how form met function. "Then we had to allow them to navigate in their environment, stop the sub from falling apart, and get them back up."

Tom Blades, the chief electrician who arguably worked longer hours than anyone, which was really saying something, detailed the challenges powering the sub with pressure-tolerant electronics, and enabling it to communicate with the surface. "None these parts existed, they all had to be made and pressure-tested," he explained. With typical understatement, he added: "It had its challenging moments."

The hydrographer, Cassie Bongiovanni, described the Kongsberg EM 124 sonar, an unsung hero of the expedition. With unprecedented precision, the sonar team had mapped 688,160 square kilometers, including 450,160 square kilometers that had never been mapped, and identified over 100 new seafloor features, which the expedition would be allowed to name. In total, the expedition examined the geomorphology of ten subduction trenches, two fracture zones, and one mid-ocean ridge basin in detail. It found previously unknown topography, like sea mounts that cause tsunamis when they are broken by shifting subduction plates and jammed back into the Earth's crust. The mapping would allow scientists who did future mapping after a seismic event to go back and see what had changed. All of the data was being given to GEBCO (General Bathymetric Chart of the Oceans) for its initiative to map 100 percent of the seafloor in high resolution by 2030.

Though the scientific part of the expedition didn't meet all of Jamieson's somewhat lofty objectives, he told the crowd that a great deal had been accomplished. The landers and sub shot more than 500 hours of video, capturing the larger, mobile animals in the deepest parts of each trench, allowing him to compare their biology and behaviors. He had found endemic Hadal species isolated in every trench. He was also able to take a biodiversity census in the Abyssal Zone, the area from 3,600 to 6,000 meters that comprises half of the Earth, and determine what was a local phenomenon and what was a global one. He could now study behavioral and biological adaptations of life in this zone.

The landers had collected 400,000 biological samples and 1.5 million meters of water data, as well as bottom water at every deep. About forty new species, including six species of fish, were discovered. At the Hadal depths, Jamieson observed bacterial mats that do not use energy from the sun, but rather use energy from chemicals or minerals derived from the seafloor. He and his team, he said, would be busy for years analyzing all the data they had captured.

"We will eventually be in a position to make a big statement about how the deep sea is functioning," Jamieson said. "But this is not necessarily an endeavor to do with studying deep sea animals. This has to do with evolution of life on Earth and the rules that govern life on Earth."

The collective impact of all the finds would help continue to tell the story of the interconnectivity of the oceans. In his presentation, Geffen talked about the potential impact of the ambitious five-part Discovery Channel series, which would be broadcast worldwide at a date to be determined. His hope was that it would inspire people to turn their attention to the oceans.

"For the first time, I think it's going to make people not look up at the stars, but look down at the ocean and realize that the deep ocean is just as important as the planets are," he said.

During the question and answer period that followed the screening of a teaser reel for the series, Vescovo was asked about the difference in the challenge of climbing Mount Everest and diving the Challenger Deep, a dual feat that now recorded him as the only person who had gone from the highest point on Earth to the lowest, a separation of some 19,773 meters, 64,872 feet, or 12.24 miles.

"Being at the top of Everest is physically and mentally punishing," he said. "It is a beating you take for two months to climb that mountain.

There is a lot of planning, but you also have to have adaptability. The Five Deeps is much more of an intellectual challenge. The organizational things you have to do to make this happen, the technical adaptability, are complex. So they are very different challenges, but the sense of accomplishment is very similar."

As McCallum wound down the program, he wryly noted that Vescovo would be given the last word on the expedition "because he paid for it."

Vescovo gathered his thoughts quickly. First, he pointed to Don Walsh, who was seated in the first row, and asked the crowd to salute him as the first person to the bottom of the ocean and the man who "showed the way." Walsh received a standing ovation. After the guests sat back down, Vescovo decided to deliver a message that was more universal than the records, the firsts, and the discoveries, one that summed up what he hoped would make a lasting impact on anyone who became familiar with the Five Deeps Expedition.

"Unless the laws of physics forbid it, very little is truly impossible," he said. "It might be harder. It might take longer. It might be more expensive, but if people put their minds to it they can do extraordinary things. If there is one thing I've seen about individuals it is that they constantly sell themselves short. They never really achieve their full potential because they never really admit how much they can achieve. You just have to wake up in the morning and say you expect more of yourself, and then you can break your own barriers on your way to breaking other ones."

EPILOGUE
BREAKING THE WAVES

The ocean research community in the Western World is composed of a relatively small cadre of experts, certainly compared to space exploration, and everyone knows, or knows of, everyone else. The manned submersible community is even smaller. While many of the scientists, submersible builders, ship captains, and marine explorers have worked with one another at various times, there is little cohesion among them and oftentimes no small amount of competition. As for nations coming together and aiding them—as much as politicians love to publicly proclaim that they support scientific research of the oceans—the approach is scattershot and obtaining cooperation can be frustrating at best, nearly impossible at worst.

Victor Vescovo entered that community with a bang by commissioning the *Limiting Factor*, forming the Five Deeps Expedition with a team comprised of the best of the best, and then successfully diving to the bottom of the five oceans and conducting multiple facets of scientific research. He and his team found that some of the toughest challenges were dealing with people in the oceanographic research world and also with the bureaucracy of foreign governments that either don't have a vision for greater ocean research, or reflexively say "No" to any requests simply because it is easier and less risky. Ironically, by the end of the expedition, the actual act of diving to the bottom of the oceans—which had never been done—had become the *easy* part.

Niggling differences within the ocean research community also threaten to obstruct progress. Nowhere was this more evident than in the relationship that Vescovo struck up with James Cameron, the only other person to dive solo to the bottom of the Challenger Deep and one of the highest profile stalwarts of ocean research.

~

Vescovo had contacted Cameron through Triton's Patrick Lahey. Cameron offered the newcomer guidance, but the relationship frosted over after Vescovo announced that he had broken Cameron's depth record at the Challenger Deep and then later completed the first solo dive to the *Titanic*, with cameras that would be the first to capture 4K images, no less. Things came to a head, coincidentally, the week after the Five Deeps Expedition's September 9, 2019, finale program at the Royal Geographical Society in London.

Two days after the London event, Vescovo flew to Monaco to meet with HSH Prince Albert II, a major ocean advocate. The expedition had collected arthropods at the Challenger Deep that were named after the research ship of the prince's great-great-grandfather. Vescovo presented those to Prince Albert II for placement in the Oceanographic Museum of Monaco. He also discussed taking the prince on a dive in the Calypso Deep, the deepest part of the Mediterranean Sea.

While in Monaco, Vescovo received a call from *New York Times* reporter William Broad asking him to comment on Cameron's assertion that Vescovo did not dive deeper than him in the Challenger Deep. Broad explained to Vescovo that Cameron had asked the reporter for an interview through an email from his assistant under the subject headline: "Request to speak."

Vescovo wasn't altogether surprised. Before the expedition's dives at the *Titanic* wreck, Cameron had sent Vescovo a lengthy email, one of several in their months-long, on-again, off-again correspondence. In the email, Cameron expressed his respect for Vescovo's accomplishments but disputed that Vescovo dove deeper than he did. Cameron wrote that when he dove, he didn't observe any variation in the terrain greater than one percent, and called the Eastern Pool at the Challenger Deep "essentially flat as a billiard table," which would become his standard line in the media. "You don't need to claim something that's not correct or ethical," Cameron wrote. "And I don't want to be in the awkward position to have to refute it, and throw shade on your work. It's truly been brilliant. But I must contest this one claim, and I will do so, when asked, which I will be—and I will have my facts and my sources readily at hand when I do." He concluded by saying that "we are on a collision course to be at public odds over this, unless you modulate your assertions with respect to Challenger," and asked Vescovo to please consider doing so.

Though Vescovo had prepared a comprehensive response based on the scientific data collected, he had not sent it to Cameron for fear of inflaming the situation. Vescovo was also surprised because in their very first communication with each other, Cameron had told him that at the Challenger Deep he had reached a maximum depth of "35,787 [feet, or 10,908 meters] plus or minus 5 meters. So that's the depth to beat." Why, Vescovo wondered, would he have said that if he believed it to be flat? The depth that Cameron reached was vetted and published by *National Geographic* in March 2012.

To the *New York Times* reporter, Vescovo explained that across multiple dives, with three landers, and backed up by sonar—none of which Cameron had—he had reported the depth he had dived in an area that was confirmed by multiple sonar sweeps and the sub's own onboard instruments to *not* be completely flat. Vescovo felt it was odd that Cameron was essentially asking him to argue a viewpoint that wasn't supported by his own (Vescovo's) data. Chief scientist Alan Jamieson, hydrographer Cassie Bongiovanni, Triton's Patrick Lahey, who had executed thousands of submarine dives including two to the Challenger Deep, all told Vescovo that they disagreed with Cameron's viewpoint and urged him not to cave just because it was James Cameron.

On September 16, a *New York Times* story ran with the headline: "So You Think You Dove the Deepest? James Cameron Doesn't," and subheadline: "Victor Vescovo claims to have set the record for the deepest ocean descent by a human. The director of *Titanic* demands to differ."

The *Times* story outlined Cameron's claim that the Challenger Deep was completely flat. Cameron told the reporter that he saw no variation in the terrain during his single dive. Broad also quoted Woods Hole Oceanographic Institution engineer Andy Bowen, who had sent the *Nereus* ROV down to measure the depth in 2009, three years before Cameron's expedition. "It was like being on the Bonneville Salt Flats," Bowen told the reporter, referring to the Utah desert known for its extreme flatness. When asked if he thought Vescovo's team had found a deeper point, Bowen replied: "Does it seem probable? Not really."

The story pointed out that the official Five Deeps Expedition press release listed the maximum depth Vescovo reached as 10,928 meters, but said in a footnote that the team might revise that figure in the future. It then noted that the figure had since been revised to 35,840 feet, or thirteen feet shallower at 10,925 meters. Vescovo said he went with the slightly shallower number "because it gave us the smallest measurement error term, and frankly, was the more conservative number."

In questioning one another, both Cameron and Vescovo also offered each other public kudos. Cameron called what Vescovo had done diving the five deeps "quite remarkable," but added: "Where I take exception is his saying he went deeper." In turn, Vescovo said: "I have enormous respect for him. On this point, however, I just scientifically disagree."

Scientific data, in fact, was the backbone of Vescovo's depth reading.

Because Cameron's basic premise was that the entire Eastern Pool of the Challenger Deep was flat, then it followed that anyone who went to its bottom would be at same depth. This did not apply to Walsh and Jacques Piccard's 1960 dive, as they dived in the Western Pool, which sonar data has showed was shallower, but just by two meters. Cameron had deployed his submersible and a lander to the Western Pool first, unmanned, and it reportedly bottomed out significantly shallower than the *Trieste* had reported. This prompted Cameron to dive in the Eastern Pool instead.

From a data point of view, Cameron's maximum final depth was 10,908 meters. That was based on the *Nereus* data and the CTD (Conductivity, Temperature, and Depth) data from his *Deepsea Challenger* on one dive. He cited evidence that the *Nereus* dived to 10,902 meters in his same general location for over ten hours in 2009 and did not see depth variation greater than one meter. However, at these extreme depths, the CTD sensors have uncertainties of ten to fifteen meters. Also, the science behind the pressure to depth conversion might not be 100 percent accurate in these depths as it becomes less precise the deeper you go. "There is no instrument that is that accurate with these kinds of depths," explains Bongiovanni. "It's not like it is on land where we get precision of heights when you're climbing."

Cameron's expedition did not have a sonar, and therefore relied on earlier, available maps. The expedition also did not apparently send multiple landers to the bottom of the Eastern Pool as Vescovo had done, who used not one but three of them to triangulate the sub's position with the ship on the surface whose precise GPS-derived position was known, and provided additional independent depth measurements.

Cameron's argument that Vescovo went to the same depth was based on a series of assumptions. First, it assumed that the *entire* Eastern Pool, which is the size of 2,000 football fields, has no variation at all beyond a one percent slope. It further assumed that both men dived in the area that Cameron claimed was perfectly flat. However, if either of his assumptions were incorrect—that it wasn't flat or Vescovo dived outside the "flat" area—then the possibility existed that Vescovo could have explored a deeper location.

The Mariana Trench, and specifically the Challenger Deep, is one of the most well mapped places in the oceans. In 2016, one of the last Law of the Sea cruises was run to collect bathymetric data over the Mariana Trench. This data was collected by some of the most prominent mappers in the world and went under extreme scrutiny before being published, making it some of the most accurate data collected to date. The data collected by the Five Deeps Expedition over the same area had an average difference of only two meters from that report. Both showed that the Eastern Pool was not completely flat.

It follows that Cameron was apparently discounting the fact that there could be a gentle slope or even a relatively small depression in a deep ocean trench—which is exactly what the expedition's maps had shown and also geologically quite possible, according to geologist Patty Fryer, who has extensively studied the area and was on both expeditions. Five other bathymetric maps of the Eastern Pool on other expeditions had also shown statistically significant variations in depth.

Vescovo's data was verified across multiple sources on various platforms from three separate dives. The first source was the Kongsberg EM 124 sonar, the most advanced of its kind. It collected data in all three "pools" of the Challenger Deep, and a portion of the Eastern Pool registered deepest. The second source was the landers deployed and their CTD data. The third source was the CTD data from the *Limiting Factor* itself. The data collected by the sonar measured within 1 to 3 meters of what was recorded by the depth sensors on the sub and the landers. Based on an analysis of the data, the independent nautical certifying agency DNV-GL—which had been on board for all of Vescovo's dives—certified a depth reference datum of 10,925 meters +/- 4 meters.

Tom Blades, the expedition's electrical engineer, did multiple calculations right after the dives and concluded that the final depth number was 10,928 meters, the initially reported figure. "The 10,927-meter reading at *Skaff* was verified by DNV-GL using data from 3 sensors—plus two sets of readings from the sub from two different dives—is the most accurate depth reading that we have," he wrote in an email to Vescovo. "We have established that you went a meter deeper while roaming around, and although we don't have enough readings to establish an absolute depth at that point to the same degree of accuracy, we can be certain that the relative accuracy compared to the 10,927-meter datum at *Skaff* is correct."

Months later, the expedition revised its 10,928-meter maximum number to a slightly lower 10,925 meters +/- 4 meters, after a review of the measurements to reflect the most accurate and conservative final depth reading based on all the data. At 10,921 meters, the shallower end of the range, it was still greater than the deeper end of Cameron's range at 10,913 meters. According to their error terms, Vescovo calculated that the statistical likelihood that the depths overlapped was less than 6 percent based on Cameron's own admission that his error term was +/- 5 meters. Anyone, he claimed, could do the math with what each side had published. He was also skeptical of Cameron's assertion that not just his, but *all* sonar surveys of the Challenger Deep—all of which showed depth variation to one degree or another—were unanimously wrong.

He also told others that he thought maybe it was as simple a situation as Cameron's sub having drifted on the surface or during the descent and diving farther west, or east, of the *Limiting Factor*'s dive, where sonar indicated it was a bit shallower and flatter. The same thing, he reasoned, may have happened at the Western Pool, and that would explain why the depth they registered there was so much shallower than what the *Trieste* had reported. To Vescovo's frustration, Cameron never seemed to even allow for the *possibility* that he or the *Nereus* could have drifted to a more shallow area of the Deep on his singular dive, in weather conditions that were far rougher than what the Five Deeps expedition had dived in.

"I feel that Jim is trying to pressure me into repudiating our extensive scientific data and say that our data is wrong and his is right, or he wants me to make his case for him that his data was all wrong and that he dove what we dove," Vescovo said after the *Times* article appeared. "The only problem I have with this is that it is not the most scientifically defensible position, at least, from where I'm sitting."

Lahey, who has worked with Cameron, concurred with the scientific data and also saw the terrain with his own eyes on two lengthy dives. "I spent two dives down there driving all over hell's half acre, and it's not flat like a billiard table," he said. "There is no question Victor went deeper. The fact is that Jim was on the lip of the bowl and Victor was in the center of the bowl."

McCallum, who worked on both men's deep dive expeditions, pointed out that there were egos involved. However, he was surprised that Cameron would outright reject Vescovo's scientific data, particularly since so many of same personnel were on both expeditions and could attest that the Five Deeps Expedition had far more, and more precise, calculations.

"Jim is a really clever guy and he is an intelligent human being, but he is just completely in the wrong," McCallum said. "He did ocean exploration a great disservice by getting petty. He's either not getting all of the information that he should have before he says outrageous things, or he's under a lot pressure filming *Avatar*. I'd be really curious why he is doing this."

Cameron also proactively contacted other writers who cover ocean exploration and asked them to run stories citing his claim that Vescovo didn't dive deeper and that they dove to the same depth. Most found it hard to ignore the high-profile, Oscar-winning filmmaker—though it didn't go unnoticed that Cameron had lost two box office records that same summer when *Avengers: Endgame* passed his two films, second-place *Titanic* and then first-place *Avatar*, to become the all-time highest-grossing film.

Popular Science and *Wired* posted stories on their websites. Both had sent reporters on legs of the Five Deeps Expedition, though the follow-ups were written by different reporters. In both stories, Cameron and Vescovo continued to disagree over the depth figures—and to declare their mutual admiration for one another. Even the website Watchpro.com covered the debate under the headline: "Rolex and Omega sucked into submarine battle seven miles under the ocean." After all, Omega had taken out a series of full-page newspaper ads touting Vescovo's "world record" dive, while Rolex had been a sponsor of Cameron's expedition.

A send-up article on all the coverage appeared in *New York* magazine's "Vulture" column. "This random feud between James Cameron and a diver—ahem, sorry, *superrich adventurer*—is unmissable," Hunter Harris sardonically wrote. "It has drama! Science! Emails! I don't think we've taken enough time for it. We live in a society where two incredibly rich men are fussing over their mutual hobby in the pages of the *New York Times*."

After laying out their cases and poking fun at both Cameron and Vescovo, she ultimately concluded that "between a man making 19 more *Avatar* movies and a man not making 19 more *Avatar* movies, I must choose the dark-sided, chaotic neutral of James Cameron."

Vescovo read the piece and immediately emailed Harris. The two then had a very friendly call in which Vescovo outlined his rationale for the depth he dived. In an update to her story, Harris wrote that she found Vescovo "delightfully jovial." She noted that at one point, Vescovo praised Cameron's work saying, "I love the *Terminator* movies. I love *Avatar*. I watched *Titanic* at the *Titanic*!" Harris declared that she was switching her allegiance to "our new favorite diver, Renaissance man, and Donald

Sutherland look-alike. Until I get a call from James Cameron, we are team Vescovo, baby!"

What Cameron didn't know was that Guinness World Records had asked Vescovo to come to their London office and meet with two of their editors to discuss the controversy, which, in 2020, certified and published the feat as the "Deepest Dive by a Crewed Vessel."

More significantly, the dust-up over who went deeper could actually have some reverberations for oceanographic research. In the *Wired* article, WHOI's Bowen, who had worked with Cameron and challenged Vescovo's team's claims of finding a deeper point, tried to redirect the conversation. "Whether each of those dives achieved the record-setting number, I think, is far less important than the idea that humans have now demonstrated their ability to reach any part of the ocean with a submersible that can bring back real, meaningful improvements in our understanding of the deep ocean," Bowen told *Wired*.

Of course, that was the salient point. The deep ocean community is small and unification remains essential to progress. But with two of the most accomplished ocean explorers at loggerheads, there was a possibility that the real loser of the debate would be ocean exploration itself.

Without a buyer for the Triton Hadal Exploration System—Vescovo had not heard anything further from the OceanX team and McCallum's talks with the billionaire Forrest weren't making any progress—Vescovo decided to mount a series of new expeditions in 2020 under the "Caladan Oceanic" banner, his oceanographic research company. The new dives would conduct science in previously unexplored areas and examine historic sites to enhance the field of knowledge about them, take other explorers and adventurers down with him and continue to shine a spotlight on deep ocean exploration. Among the stops was a planned return to the Challenger Deep for a series of dives to collect, in Vescovo's words, "a mountain of additional data" to settle the depth dispute "the way good scientists do." In sum, the anticipation was that these dives would further deep sea exploration in places seldom, if ever, visited and extend the *Limiting Factor*'s search capabilities. But as the Five Deeps Expedition had repeatedly encountered, getting permission to look for things in the water turned out to be far tougher than looking for them on land, or even in space for that matter.

To commemorate the anniversary of the end of World War II, Vescovo announced plans to dive the USS *Indianapolis*, the heavy cruiser sunk by the Japanese Navy, and what was believed to be the USS *Johnston*, the World War II destroyer that sank in battle in 1944. There had never been a manned submersible dive to either wreck. It seemed appropriate, given that Vescovo was a former U.S. Navy Reserve officer and he would turn over all the footage gratis to the U.S. Navy, but there was immediate pushback. U.S. Navy Heritage Command indicated that some family members of those deceased on the *Indianapolis* objected, and therefore it refused to provide the coordinates. Vescovo was surprised by the resistance, and since he stated he would never dive a wreck without the full support of its stakeholders, he shelved the plan.

A similar situation occurred with the *Johnston*. The wreck had been discovered by the private research firm Vulcan, funded by the late billionaire Paul Allen. Vulcan too refused to give Vescovo the coordinates. U.S. Navy Heritage declined to intervene to convince Vulcan to allow further examination of the wreck, odd given that the *Petrel* ROV that discovered the wreck could only reach 6,000 meters and the wreck was located at 6,200 meters—the deepest wreck reportedly ever discovered.

"I have the only sub capable of investigating the *Indianapolis* and the *Johnston*, and I don't even know where they are—and nobody will tell me," Vescovo said. "I want to go to these to provide a historical record and answer any lingering questions. But everybody views these wrecks as their possession. I mean, these are wrecks on the bottom of the ocean for heaven's sake."

He turned his focus to other areas of interest and settled on a five-phase expedition. The dives would be at the *Minerve*, a French sub that was mysteriously lost in 1968 and whose government *did* support a visit, the Calypso Deep in the Mediterranean, the Red Sea off the coast of Saudi Arabia, multiple locations in the Indian Ocean in partnership with the Nekton organization, and the Challenger Deep and the surrounding

Ring of Fire, home to many deep-sea volcanoes and the site of seaquakes and massive tectonic plate movements.

Many on the original expedition team would return.

Captain Buckle and his crew had remained with the ship and on the payroll during its maintenance period in Las Palmas, Canary Islands. Lahey agreed to bring his team from Triton back to run the submersible operations. However, now that the LARS had been perfected, the Triton team would be scaled back slightly to keep costs down, though Triton would now be charging its normal rates rather than the reduced ones during the Five Deeps. Triton also needed many of its people for other projects, including building a 7,500/3 sub with an acrylic sphere. Most notably, after Lahey went to Barcelona to conduct sea trials, he stepped back from the new expedition and put the very capable Kelvin Magee in charge.

Triton had been left in a slightly unsettled position at the end of the Five Deeps Expedition. On the one hand, it had built and proven the world's first classed ultra-deep diving submersible. Triton received its full costs for building the sub and staffing the dives, and also collected $2.5 million in bonuses for completing the expedition's goals. However, because the system had not been sold, the company had not realized any other bonuses or profit from the sale of the submersible, only significant media exposure.

Bruce Jones, Lahey's partner, said that the project consumed a great deal of the company's manpower and attention, including all of its senior employees, but hadn't yielded the hoped-for financial benefits.

"At the time we agreed to the deal based on the best information we had," Jones said. "We thought it was workable, and we are grateful to Victor for being our client. But he's a pretty shrewd negotiator. I regretted some of the terms that we agreed to. I look at it from a different perspective than everybody else. What are the dollars and cents? How does this advance the company? More people might know what Triton is, but that's not really marketing, that's entertainment."

Jones, of course, did not set foot on the ship during the expedition. Vescovo invited him several times, but after his son was dismissed and his wife was denied going on the company's discretionary dive at the

Challenger Deep, Jones did not feel welcome. When Triton entered the agreement with Vescovo, Lahey and Jones were splitting the supervision of the project, with Lahey overseeing the design and build of the sub and Jones taking on the refit of the ship and planning of the expedition. But things unraveled when the cost of the refit soared, and it was later agreed that Jones would not lead the expedition. The entire project had gone from being the ultimate project in his 34-year career to a business arrangement that he was trying to maximize.

Toward the end of the expedition, in the summer of 2019, Triton hired an adviser and put itself up for sale. Jones planned to cash out entirely while Lahey would retain his stake in the company. "After forty years of being self-employed, I've had enough," Jones said.

For his part, Lahey remained optimistic about selling the Hadal Exploration System. "My mother always said, 'Take off those rose-colored glasses,' but I can't," he said. "The fact is I was absolutely certain, as Bruce was, that the system would be sold before Victor even finished the Five Deeps because of how unique this craft is. It is a unicorn. If it's not sold, it would be devastating for me, but it will be sold because there is nothing like it in the world. Yes, we took a risk. But as a small business owner in the business that I am in, everything that I do is a risk. I'm so used to pushing all my fucking chips to the center of the table, and I do it all the time."

Jones and Lahey felt that Triton was owed a large payment under its arrangement for Vescovo continuing to dive the *Limiting Factor* after the Five Deeps Expedition, but Vescovo countered that continuing to dive and perfect the sub was the best way to keep marketing it. After some back and forth, the Triton duo agreed.

For media on the 2020 expedition, Vescovo planned to use a scaled-down film crew and post the videos on YouTube and the Caladan website, as well as having his PR firm send them to interested media outlets. He was also discouraged that the five-part Discovery Channel series had not aired. (As of August 2020, the one-year anniversary of the completion of the Five Deeps Expedition, the series had still not been scheduled, due in part to complications brought on by the COVID-19 pandemic. Vescovo was told by his contact at Discovery that the series would definitely air and that the network just wanted to make sure it had the best launch slot.)

The Five Deeps Expedition science team would largely remain intact. Hydrographer Cassie Bongiovanni would continue the extensive mapping effort at the new locations and during many of the transits. Although Dr. Alan Jamieson looked back on the expedition with a tinge of disappointment, he welcomed the chance to return as chief scientist, as the new planned dives had many unexplored features. His feeling was that with the sub now functioning and the LARS fine-tuned, the new expedition could undertake more science and discovery missions.

"A few months after the Five Deeps ended, I began to see it with different eyes," Jamieson said. "The original plan was too ambitious. I guess I was never really told the whole thing would essentially be a trial cruise. When I see the video from 103 landers and the specimens from these dives, it is truly amazing to have done all that from so many places in a year. Likewise the maps from the multibeam are amazing, too—650,000 square kilometers (about 251,000 square miles) of footage. While the Java, Sirena, and Molloy sub videos are great, it is disappointing or respectfully, frustrating, when I think about how much work went into the sub to get the Five Deeps done to walk away with so little useable footage. But now the sub is working, perhaps the next year or so will be very different."

After two days of sea trials off the coast of Barcelona in late January 2020 to ensure the *Limiting Factor* was in full working order after several system updates, the *Pressure Drop* sailed to Toulon, France, to pick up the team for the first phase of the Caladan 2020 Expeditions: a dive to the *Minerve*. When the French submarine was discovered in July of 2019, the *Minerve*, which sunk in 1968, was the last missing submarine operated by a Western nation from a decade that had seen several sub disasters. The dives would be the first in a manned submersible to the remains of the submarine.

P. H. Nargeolet, who is highly regarded in the French maritime community and has done numerous dives for the government, was able to secure the permit. Even though searching for clues of what had happened was of interest to family members of the deceased, the French government was reluctant to allow the dive. Nargeolet worked behind the scenes with the families who were looking for the definitive reason of why the sub sank, and also to achieve closure. Eventually, the French government and navy agreed.

On the first dive, Vescovo took retired French rear admiral Jean-Louis Barbier down to the wreck, which lies at a depth of 2,350 meters (7,710 feet) and is scattered across a square mile of seafloor. The two spent over three hours conducting a comprehensive examination of the wreck. On the way up, Barbier told Vescovo that he was now pretty sure what had caused the vessel to sink.

Based on his direct visual observation of distressed metal, Barbier posited that the submarine had been run over by a cargo ship in high seas. In a freak accident, the ship, his theory went, did not see the sub's snorkel above the water, and the sub didn't detect the ship. When the ship plowed over the snorkel, it snapped, sending water rushing into the sub.

Barbier planned to review the footage from the external high-definition cameras on the *Limiting Factor* and write a formal report for the French government with his conclusions.

On the second dive, Vescovo took Hervé Fauve, the son of the submarine's captain, to place a stone plaque on the *Minerve*, as the manipulator arm was finally in full working order. Fauve grew emotional seeing where his father had died up close, and Vescovo surprised him by playing "La Marseillaise" on his phone as they laid the memorial. This highly personal moment for someone's life who had been forever altered by a deep sea tragedy was something that the *Limiting Factor* could provide for any family member who had lost someone in a sunken vessel, anywhere in the world.

The second phase of Caladan's 2020 expedition was a dive at the Calypso Deep, located in the Hellenic Trench, the deepest stretch of the Mediterranean Sea. The *Pressure Drop* ported in Kalamata, Greece, to pick up HSH Prince Albert II of Monaco for the dive. This permit was issued rather quickly in consideration of the special guest's relationship with the Greek government, a welcome development to say the least. It was beginning to feel like political connections were needed to conduct important research at sea.

Prince Albert II's great-great-grandfather had been a champion of ocean research, and his father had inaugurated the Oceanographic Museum of Monaco. The museum houses the laboratory of Prince Albert I's research yacht, on which observations on the understanding of anaphylaxis, the potentially life-threatening allergic reaction, were made, leading to a Nobel Prize for Dr. Charles Richet. Prince Albert II had taken up the fight to save the oceans, advocating for protection of marine biodiversity and prevention of overfishing.

On the dive to this, the deepest point in the Mediterranean (5,109 meters, or 16,762 feet), Vescovo and Prince Albert II saw a fairly active marine life population, but also a disturbing amount of garbage on the bottom, including a Coke can, several plastic bags, and what appeared to be a discarded garden hose. Prince Albert II, who called the experience of seeing the seafloor up close "amazing, just amazing," but was admittedly disturbed by the extent of the contamination, became the deepest diving head of state in the world. Doing such a dive himself would add to his international platform to advocate against human contamination in the seas.

The third phase was a partnership with the King Abdullah University of Science and Technology (KAUST) in Saudi Arabia to dive in the Red Sea, an inlet of the Indian Ocean bordering Africa and Asia. Vescovo had met the university's director, Justin Lee Mynar, at a conference in Northern California sponsored by the Schmidt Ocean Institute. A fellow Texan and ocean enthusiast, Mynar jumped at the chance to have his scientists dive the Red Sea with Vescovo and his team. Two KAUST scientists came aboard for the first manned dives to the bottom of the Red Sea.

On the first dive, Vescovo and Jamieson dived to the bottom of the Kebrit Deep (*kebrit* is Arabic for sulfur) at 1,470 meters (4,823 feet). They spent five hours exploring the "brine pools" on the seafloor. These milky areas at the bottom are very different in character from the seawater above them. Because the salinity is up to eight times greater than water, there is an underwater "shoreline" between the regular sea above and the pools. For Jamieson, it was a chance to recover water samples and analyze them as part of his global look at salinity in the seas and the sustainability of life in varying undersea climates.

Vescovo then did two dives with university scientists in the Suakin Trough, the deepest area of the Red Sea. On the first dive in the trough, he drove into an inactive underwater volcano. The second dive went to the deepest point at 2,772 meters (9,111 feet), shallow compared to the deep oceans. No large rivers drain into the Red Sea; two deserts border the sea, and the resulting high evaporation rates create the large brine pools, or lakes that, like the Kebrit Deep, the Suakin Trough also contains.

On that dive, Vescovo was originally supposed to take down a member of the Saudi royal family to the bottom, but those plans fell through for various reasons so he drafted a young Saudi ROV engineer on board to go down instead. Mohammed A. Aljahdli became the first person, with Vescovo, to the bottom of the Red Sea, and, they believed, Aljahdli became the deepest-diving Saudi in history. The twentysomething graduate student couldn't believe his luck.

But, as it so often happens in deep-sea exploration, a political twist intervened. A Saudi commodore was on board the ship to oversee the dives. It became apparent that he was part of the traditional faction in Saudi Arabia that was feeling upstaged by the growing power of the country's modernizing faction, of which KAUST was a part, and he took advantage of being "in charge." The commodore went from being a minor annoyance, needing his cabin changed and demanding special meal times, to a problem. At one point, he insisted that Vescovo immediately wire an $8,000 fee to his personal bank account, which he said was proper compensation for his supervisory duties, and to provide proof of payment the very next day.

After the dive to the bottom of the Red Sea had been completed, the commodore inserted himself in the photograph of Vescovo and the young Saudi holding aloft the Saudi flag. The photo op completed, the commodore decided he had had enough of being at sea.

Within hours, he ordered the ship to sail to port—before the landers were even recovered. Quite obviously, Jamieson and the captain on this leg, Alan Dankool, refused to leave hundreds of thousands of dollars of equipment behind. Additionally, there was an important science dive set for the following day.

In a tense meeting on the bridge, Vescovo appealed to the commodore that the permit was for one more dive, but the man wouldn't budge. Just as Vescovo had avoided any possible ramifications from a confrontation with the Indonesian government by not returning to the country after diving the Indian Ocean without permission, he did not want to alienate a member of the Saudi Ministry of Defense who could make trouble for the ship leaving the Red Sea. After the landers were secured, Vescovo told the captain to sail for port, thereby scrapping the final dive. Jamieson was incensed, but Vescovo was resigned to how things worked in the Middle East. "Just when you think things are going well," he told the captain, "they don't."

Later that evening, the commodore stalked through the control room and demanded that the expedition photographer and the sonar operator turn over originals of all the videos, photos, and bathymetric maps. He told them Vescovo would have to sign a document confirming all copies were destroyed. None of this demand was part of the original arrangement made through KAUST, so Vescovo did not comply and refused to sign any such document.

En route to port in Jeddah that night, Vescovo spoke to KAUST's Mynar, who was horrified at what had happened and apologized profusely. But there was little that could be done. The following morning around 4:30 A.M., the ship tied up in the deserted port. Vescovo, deliberately avoiding any contact with the commodore who was sleeping in his cabin, quietly walked off the ship. An apologetic KAUST official drove him to the airport and ensured he cleared outbound customs and got on his plane, all before the commodore, apparently, woke up.

The *Pressure Drop* left the Red Sea without incident, and headed for the Seychelles. For the fourth phase of the 2020 expedition Vescovo had agreed to lend the system to Oliver Steeds and his Nekton exploration mission at cost. Steeds and Vescovo had talked early on about doing something together, but Steeds was busy with his own ocean awareness program and Vescovo wanted to complete the Five Deeps. Both Lahey and McCallum had worked with Steeds and continually encouraged Vescovo to talk further to him, which had resulted in the collaboration.

During the Five Deeps Expedition, Steeds had run a series of shallow-water submersible dives called First Descent in collaboration with the University of Oxford, using Triton-built submersibles. Designed to call attention to the protection of the world's seas, Steeds did a shallow-water dive to 400 feet with Seychelles president Danny Faure that was covered by the *New York Times*.

To increase visibility for its cause and to execute science dives to study preservation, Nekton planned a monthlong exploration of the Indian Ocean in March of 2020 using the Triton Hadal Exploration System. This was the type of expedition that Jamieson had talked about using the *Limiting Factor* for—multiple dives in deep water within a single area to paint a more detailed ecological picture. But growing concern over the rapid spread of the COVID-19 virus canceled the expedition at the very last minute. Both sides hope to reboot the expedition in the future—the upside being that two different groups involved in ocean research had agreed to deeply collaborate. That alone was progress in ocean research.

~

The ship returned to the Challenger Deep in June of 2020 for a series of dives dubbed the "Ring of Fire Expedition" with notable figures in their worlds and also a couple of paid dives with wealthy adventurers. To continue to fund the operation of the Hadal Exploration System without a partner, Vescovo had turned to offering dives to the Challenger Deep priced at $750,000, not unlike the space tourism industry that was coming online in 2021 at far higher prices. The capital was used to help pay for two months of seabed mapping for NOAA and a separate mapping project for the International Hydrographic Organization.

For the first dive, Vescovo invited astronaut Dr. Kathy Sullivan, who holds a PhD in marine geology and was the first American woman to walk in space, as his guest, to shine a further light on the work the new expedition was doing. They dived the Eastern Pool for an hour and a half and she became the first *person* to walk in space and reach the bottom of the ocean.

After returning to the ship, the pair received a call from the SpaceX *Crew Dragon*'s astronauts Chris Cassidy, Ben Behnken, and Doug Hurley, who were docked at the International Space Station, around 254 miles above the Earth. "Congratulations on your scientific accomplishments in the deep underwater, as we're trying to accomplish the same on the International Space Station," Cassidy said.

On the call, Cassidy, Sullivan, and Vescovo discussed the similarities in the two vessels, the *Crew Dragon* and the *Limiting Factor*, namely that they were designed to be reusable and make multiple trips to their respective destinations. Sullivan remarked that the *Limiting Factor* was "like having a craft that can make daily trips to the Moon." Vescovo also told Cassidy that if his boss—SpaceX founder and CEO Elon Musk—would give him a ride in "his," he would give Musk a ride in "mine." It was a proposal that Vescovo had made previously in public and hoped would one day come to pass.

For Sullivan, the dive and phone call were highlights in a career filled with them. "As a hybrid oceanographer and astronaut, this was an extraordinary day, a once-in-a-lifetime day, seeing the moonscape of the Challenger Deep and then comparing notes with my colleagues on the ISS about our remarkable reusable inner-space outer-space vehicles," Sullivan said.

Vescovo invited Kelly Walsh, the son of *Trieste* captain Don Walsh, to retrace his father's pathbreaking 1960 dive in the Western Pool. Walsh dived forty-six feet deeper than his dad and stayed on the bottom nearly ten times longer, thereby officially joining the "family" business.

On a nearly four-hour dive, the two saw what they believed to be a discouraging amount of fine, white tether cable, composed of fiber optic cable and plastic. The tether, which is attached to ROVs that dive the ocean floor, had apparently been cut and left to sink to the bottom, where it would remain forever.

"The most amount of human contamination I've seen by far at the Challenger Deep was left by scientists, or maybe military operators," Vescovo said.

In an effort to open a further door with the Woods Hole Oceanographic Institution, Vescovo invited Dr. Ying-Tsong Lin, a deep water acoustics expert, to dive the Central Pool with him. Lin, who was born in Taiwan, became the first person of Asian descent to visit the ocean floor. It was also a poke in the eye directed at China—over China's harsh treatment of Taiwan and being a nation Vescovo is no fan of—which has struggled in its quest to send a sub to the bottom of the Challenger Deep.

Though Vescovo and Lin saw no tether cable in the Central Pool, Vescovo showed Lin the images from the Western Pool. They later conferred with Casey Machado, a WHOI research engineer in the Deep Submergence Laboratory who has run ROV operations at the Challenger Deep. Machado explained that WHOI uses a more sophisticated, thinner tether system to connect to its ROVs and consequently does not use such white tether cables. Her suspicion was that the debris that Vescovo saw might be from a more rudimentary system that perhaps failed to connect multiple landers on the seafloor some distance apart, or from a less sophisticated ROV that had broken its tether. Vescovo believed that those signs possibly pointed to Chinese government or military researchers who may have been conducting operations at the Challenger Deep.

"The Chinese have admitted they have a hydrophone set or two down there, and it appears in trying to recharge or communicate with it, or them, a pretty big amount of tether cable snapped and now is just drifting down there," Vescovo said. "It's the equivalent of 'space junk' that has made parts of space very dangerous for travel. One thing is certain, if those who leave behind this kind of waste don't pay more attention to what happens to these tethers, the deeps could be become as contaminated and dangerous to dive as parts of outer space."

Of course, until now, no one other than the people who left it behind would have even known it's there.

In another first, Vanessa O'Brien, who had completed the Explorers' Grand Slam, dived with Vescovo in the Eastern Pool and became the first woman to summit the highest mountain, Mt. Everest, and reach the deep ocean floor. Rather than solicit sponsors during the COVID-19 pandemic, O'Brien paid for the dive out of her own pocket.

Two other paid dives were completed with entrepreneurs John Rost, a sort of doppelgänger to Vescovo who explores for the thrill of exploration, and Jim Wigginton, a former marine who sought to raise awareness for the Punya Thyroid Cancer Endowment Fund that he established in memory of his late wife. Vescovo and Rost stayed down for 4 hours and 7 minutes, breaking Vescovo's solo record on the bottom by 5 minutes.

In addition to the dives, the expedition did twenty-four lander deployments across the Challenger Deep. The landers and the dives verified the deepest points of each pool—the Central at 10,915 meters, the Western Pool at 10,923 meters, and the Eastern Pool at 10,925 meters, all carrying a +/- 4 meter deviation. For maximum accuracy, the expedition added battery-operated depthometers to the landers and combined that data with the CTD and sonar data. On the final dive in the Eastern Pool, the team actually placed all three CTD sensors and all three depthometers on the sub, and again the average measurement came back at 10,925 +/- 4 meters.

"I have spent twenty-four cumulative hours on the bottom of the Challenger Deep, and I can tell you that it is not flat," Vescovo says. "I saw with my own eyes bare rock with a sloping ledge in Eastern Pool near the central area. And I have 4K video of it."

The *Limiting Factor* had now taken ten people to the bottom of the Challenger Deep and changed a key statistic in the space-versus-ocean conversation. Adding to the first three divers, Don Walsh and Jacques Piccard in 1960 and James Cameron in 2012, thirteen people have now been to the bottom of the deepest ocean—one more person than has walked on the Moon.

At its core, all exploration is rooted in curiosity and wonder, be it the Mercury, Gemini, and Apollo missions to reach the Moon, the *Trieste*, *Deepsea Challenger*, and *Limiting Factor* dives to reach the bottom of the deepest ocean, or Roald Amundsen's quest to reach the South Pole. What is up there? What is down there? What is out there? And it is rooted in the desire to find out for one's self, not to wait for someone else to look, but to look yourself. And when the results are found, whether the entire world watches in awe, there is a casual fascination, or barely anyone at all notices, the feeling of the participants is the same, one of satisfaction rooted in the journey. But over time, many of these successes by explorers and their teams to deliver something for themselves has delivered something for all humankind.

The people who are so compelled must step up, assemble a team, and risk whatever is necessary—including their lives if need be—because they have to. From the personal drive of the earliest pioneers like Columbus and Magellan, who proved that the world was not flat, to the great explorers of modern times like Ernest Shackleton, Edmund Hillary, and Jacques Cousteau, who led expeditions that helped shaped our knowledge of the planet, to the space innovators like Neil Armstrong, Buzz Aldrin, and John Glenn comes progress that impacts us all.

Many far less famous people possess that similar desire, that "exploration gene," to push their own limits with groundbreaking undertakings that have the promise to find something but with no guarantee of success. Dr. Glenn Singleman, who served on both Cameron's and Vescovo's expeditions, has studied the accounts of past adventurers and present-day ones firsthand, and also knows from where he speaks—he is the world-record holder in base jumping who plummeted a breathtaking 37,650 feet from a hot-air balloon. While Singleman acknowledges that the goal of expeditions can be scientific discoveries that carry existential benefits, his perspective on first-of-their-kind voyages to uncharted territories differs from that of most journalists, scientists, and awestruck observers.

"What any adventure is about is an opportunity to take an internal journey," he says. "All the external stuff is not even relevant. It's a reminder to take the internal journey. You have to go to places inside your own psyche that most people never visit. You have to find resources in yourself that most people never find. You have to come up with options, solutions, and insights that most people never get the opportunity where they are put in the situation to develop those things, and that's the real value of doing this."

At one event, Vescovo was asked what the return on the Five Deeps Expedition was to him. Why had he invested so much of his net worth to build the world's first classed submersible and traverse the globe to dive to the deepest points of our oceans?

"It's personal, so it's incalculable," he responded without hesitation. He thought for a few seconds and then expanded on his answer. "We don't know the return. This is a door. Up until now, the deep ocean has been this huge, impenetrable mystery with heaven knows what treasures. Once every decade you get to open the door, only to see it slammed shut. What we have tried to do is build a vehicle that opens it wide so we can find out what is on the other side."

Vescovo, who derives inspiration from Amundsen and Shackleton but steers clear of any comparisons, later picked up on his own observation and reflected on how he viewed great explorers' precedent-setting feats, and how people responded to their quests:

"The Wright Brothers built an aircraft that could fly, and some people were like, 'So what? Good for them. It's not faster than a railroad. It can't carry anyone. You are lying down on canvas to even fly it. It's really cool, but . . . so what?'

"These are the kind of people who said, 'Neil Armstrong walked on the Moon, and we got a bunch of Moon rocks for like $100 billion. What does that have to do with the price of beer?'

"That's a little frustrating for me. I personally believe that triumphs like that are what it is to be human, which is to technologically push the boundaries of what we can do as a species. There are people who get it and see the technological advancement, and whether it is incredibly useful or not in the moment, it's opening a door that didn't exist.

"So the point is, we don't immediately know where some of these things are going to take us. We don't know what other discoveries will result from them. I believe the same is true of the *Limiting Factor*—that we just don't know yet."

ACKNOWLEDGMENTS

The first thing I read when I pick up a new book are the acknowledgments, which is kind of like opening the Sunday *New York Times* and reading the "Vows" section first. Alas, I notice how many writers use the space to thank not only those who helped their current pages make it into print, but people who have guided them along the way. Since I haven't yet, now I will.

The late filmmaker Anthony Minghella once told me in the wee hours of the morning while we were standing in the mountains of Transylvania in a foot of mud with wet snow falling on the gloomiest, most freezing night the planet might have ever seen, as he was holding a patient crew and some very notable actors at bay so that a backwoodsman with a three-foot beard could coax several oxen through the film set: "You must choose your traveling companions well or you will arrive in tatters." I have certainly tried to take that advice.

Let's start with this project, much of which took place on a ship that absorbed its share of swells in the middle of the world's oceans. First and foremost, thank you to Victor Vescovo. He granted me full access without any restrictions to all aspects of the Five Deeps Expedition before he knew it would work, and he never wavered, even when things weren't going well, which was often. It has been said many times by many people on the expedition that without Victor, none of this would have happened—and that certainly includes this book. Whatever his place in the history of oceanography becomes won't be my lasting memory of this project. What I take from this experience is his attitude that if you have the courage to be the first person through the wall, come what may, then you will inspire the next person.

I thank everyone on the ship who so generously gave me their time and allowed me to eavesdrop at often inconvenient times: Patrick Lahey, Rob McCallum, Anthony Geffen, Dr. Alan Jamieson, John Ramsay, Captain Stuart Buckle, Tom Blades, Jonathan Struwe, Dr. Glenn Singleman (MD), P. H. Nargeolet, Cassie Bongiovanni, John Davies, Don Walsh, Heather Stewart, Ian Syder, Kelvin Magee, Parks Stephenson, Dr. Patricia Fryer, Frank Lombardo, Steve Chappell, Colin Quigley, Karen Horlick, Tim Macdonald, Manfred Umfahrer, Marcos Benavides, Shane Engler, Hector Salvador, Mat Jordan, Paddy Russell, and Richard Varcoe, as well as those off the ship: Bruce Jones, Kyle Harris, Dick DeShazo, Matt Lipton, Enrique Alvarez, General Charles Krulak, Manny Montes, Victoria Webster, and Monika Allajbeu. Thanks also to my cabinmates on the sailings to the deeps: Alan, Anthony, Marcos, James Blake, and Josh Dean, and to the absolutely first-rate *Pressure Drop* crew for making the onboard experience less woozy and more enjoyable.

It's beginning to be an echo in my acknowledgements, but thanks once again to agent extraordinaire Andrew Stuart for hanging in there on this one and finding the right home for it.

That home is Pegasus Books, a comfortable one for an author. My editor, Jessica Case, brought the book in and took good care of it. I appreciate the efforts of the entire team: publisher Claiborne Hancock, production designer Maria Fernandez, copy editor Dan O'Connor, and cover designers David Carlson and Chris Gilbert at Gearbox.

Backing up a bit, I retain an appreciation for the two teachers who gave me the love of language, the late Gladys Preston and Paul Hamer. Making a go of it in the writing trade early on required newspaper and magazine editors who gave me a chance and consequently dealt with my sometimes gummy prose and taught me how to make it better: Neil Amdur, Michael Keating, John Phillips, Julie Ward, John Preston, Linda Lee, Bill Tonelli, Maggie Murphy and, first among equals, Elizabeth Mitchell.

I've had a long list of supporters, too long to type, but I must single out a few whose inexhaustible supply of good will and bits of carefully delivered wisdom leave me indebted to them: my mentor Mike Medavoy, one of the last class acts in the film industry (will the Academy please give him the Thalberg Award?); Sherry Lansing, who always says yes before I even ask, and the late Wayne Rogers, who instilled in me the importance of curiosity.

And finally, I raise my glass to the guys who have had my back over the years: Jon Klane, Greg Powell, Nick Reed, Rahsaan Peak, Neal Baer, Jeff Goldberg, Jim Dicker, Steve Cochrane, Rick Remmert, and Carl Berkelhammer. The next round is on me.

—Josh Young
Palm Desert, California

ADDITIONAL RESOURCES

Five Deeps Expedition: https://fivedeeps.com

Caladan Oceanic: https://caladanoceanic.com

Triton Submarines: https://tritonsubs.com

EYOS Expeditions: https://www.eyos-expeditions.com

Atlantic Productions: https://atlanticproductions.tv

INDEX